CRC SERIES IN AGRICULTURE

Editor-in-Chief

Angus A. Hanson, Ph.D.
Vice President-Research
W-L Research, Inc.
Highland, Maryland

HANDBOOK OF SOILS AND CLIMATE IN AGRICULTURE

Editor
Victor J. Kilmer (Deceased)
Chief
Soils and Fertilizer Research Branch
National Fertilizer Development Center
Tennessee Valley Authority
Muscle Shoals, Alabama

HANDBOOK OF PLANT SCIENCE IN AGRICULTURE

Editor
B. R. Christie, Ph.D.
Professor
Department of Crop Science
Ontario Agricultural College
University of Guelph
Guelph, Ontario, Canada

HANDBOOK OF PEST MANAGEMENT IN AGRICULTURE

Editor
David Pimentel, Ph.D.
Professor
Department of Entomology
New York College of Agricultural
and Life Sciences
Cornell University
Ithaca, New York

HANDBOOK OF ENGINEERING IN AGRICULTURE

Editor
R. H. Brown, Ph.D.
Chairman
Division of Agricultural Engineering
Agricultural Engineering Center
University of Georgia
Athens, Georgia

HANDBOOK OF TRANSPORTATION AND MARKETING IN AGRICULTURE

Editor
Essex E. Finney, Jr., Ph.D.
Assistant Center Director
Agricultural Research Center
U.S. Department of Agriculture
Beltsville, Maryland

HANDBOOK OF PROCESSING AND UTILIZATION IN AGRICULTURE

Editor
Ivan A. Wolff, Ph.D. (Retired)
Director
Eastern Regional Research Center
Science and Education Administration
U.S. Department of Agriculture
Philadelphia, Pennsylvania

Lethbridge Community College Library

CRC Handbook of Engineering in Agriculture

Volume II
Soil and Water Engineering

Editor
R. H. Brown, P.E.
Chairman Emeritus
Division of Agricultural Engineering
University of Georgia
Athens, Georgia

CRC Series in Agriculture
A. A. Hanson, Editor-in-Chief
Vice President-Research
W-L Research, Inc.
Highland, Maryland

CRC Press, Inc.
Boca Raton, Florida

Library of Congress Cataloging-in-Publication Data

CRC handbook of engineering in agriculture.
  (CRC series in agriculture)
  Includes bibliographies and indexes.
  1. Agricultural engineering—Handbooks, manuals,
etc. I. Brown, R. H. (Robert H.) II. Series.
S675.C73 1988   630   87-21870
ISBN 0-8493-3861-1 (v. 1)
ISBN 0-8493-3862-X (v. 2)
ISBN 0-8493-3863-8 (v. 3)

This book represents information obtained from authentic and highly regarded sources. Reprinted material is quoted with permission, and sources are indicated. A wide variety of references are listed. Every reasonable effort has been made to give reliable data and information, but the author and the publisher cannot assume responsibility for the validity of all materials or for the consequences of their use.

All rights reserved. This book, or any parts thereof, may not be reproduced in any form without written consent from the publisher.

Direct all inquiries to CRC Press, Inc., 2000 Corporate Blvd., N.W., Boca Raton, Florida, 33431.

© 1988 by CRC Press, Inc.

International Standard Book Number 0-8493-3861-1 (v. 1)
International Standard Book Number 0-8493-3862-X (v. 2)
International Standard Book Number 0-8493-3863-8 (v. 3)

Library of Congress Card Number 87-21870
Printed in the United States

EDITOR-IN-CHIEF

Angus A. Hanson, Ph.D., is Vice President-Research, W-L Research, Inc., Highland, Maryland, and has had broad experience in agricultural research and development. He is a graduate of the University of British Columbia, Vancouver, and McGill University, Quebec, and received the Ph.D. degree from the Pennsylvania State University, University Park, in 1951.

An employee of the U.S. Department of Agriculture from 1949 to 1979, Dr. Hanson worked as a Research Geneticist at University Park, Pa., 1949 to 1952, and at Beltsville, Md., serving successively as Research Leader for Grass and Turf Investigations, 1953 to 1965, Chief of the Forage and Range Research Branch, 1965 to 1972, and Director of the Beltsville Agricultural Research Center, 1972 to 1979. He has been appointed to a number of national and regional task forces charged with assessing research needs and priorities, and has participated in reviewing agricultural needs and research programs in various foreign countries. As Director at Beltsville, he was directly responsible for programs that included most dimensions of agricultural research.

In his personal research, he has emphasized the improvement of forage crops, breeding and management of turfgrasses, and the breeding of alfalfa for multiple pest resistance, persistence, quality, and sustained yield. He is the author of over 100 technical and popular articles on forage crops and turfgrasses, and has served as Editor of *Crop Science* and the *Journal of Environmental Quality*.

PREFACE

CRC SERIES IN AGRICULTURE

Agriculture, because of its pivotal role in the development of civilized societies, contributed much to the development of various scientific disciplines. Thus, agricultural pursuits led to the practical application of chemistry, and gave rise to such major disciplines as economics and statistics. The expansion of scientific frontiers, and the concomitant specilization within disciplines, has proceeded to the point where agricultural scientists classify themselves in an array of disciplines and subdisciplines, i.e., nematologist, geneticist, physicist, virologist, and so forth. Nevertheless, within the framework of these various disciplines and mission oriented agricultural research, information of primary interest and concern in the solution of agriculturally oriented problems is generated. Although some of the basic information finds its way into the plethora of reference books available within most disciplines, no attempt has been made to develop a comprehensive handbook series for the agricultural sciences.

It is recognized that there are serious difficulties in developing a meaningful handbook series in agriculture because of the range and complexity of agricultural enterprises. In fact, the single common denominator that applies to all agricultural scientists is their universal concern with at least some aspect of the production and utilization of farm products. The disciplines and resources that are called for in a specific investigation are either the same or similar to those utilized in any area of biological research, or in any one of several fields of scientific endeavor.

The sections in this handbook series reflect the input of different editors and advisory boards, and as a consequence, there is considerable variation in both the depth and coverage offered within a given area. However, an attempt has been made throughout to bring together pertinent information that will serve the needs of nonspecialists, provide a quick reference to material that might otherwise be difficult to locate, and furnish a starting point for further study.

The project was undertaken with the realization that the initial volumes in the series could have some obvious deficiencies that will necessitate subsequent revisions. In the meantime, it is felt that the primary objectives of the Section Editors and their Advisory Boards has been met in this first edition.

A. A. Hanson
Editor-in-Chief

ADVISORY BOARD

Paul Turnquist, Ph.D.
Head
Agricultural Engineering Department
Auburn University

G. O. Schwab, Ph.D.
Professor
Agricultural Engineering Department
Ohio State University
Columbus, Ohio

D. Heldman, Ph.D.
Chairman
Agricultural Engineering Department
Michigan State University
East Lansing, Michigan

F. J. Humenik, Ph.D.
Professor
Agricultural Engineering Department
North Carolina State University
Raleigh, North Carolina

H. J. Hansen, Ph.D.
Western Region
Agricultural Research Service
Oregon State University
Corvallis, Oregon

W. H. Brown, Ph.D.
Head
Agricultural Engineering Department
Louisiana State University
Baton Rouge, Louisiana

S. M. Henderson, Ph.D.
Professor
Agricultural Engineering Department
University of California
Davis, California

R. L. Pershing
Engineering Division
Advanced Production
John Deere Ottumwa Works
Ottumwa, Iowa

T. E. Stivers
The T. E. Stivers Organization, Inc.
Consultants and Engineers
Decatur, Georgia

CONTRIBUTORS

B. J. Barfield, Ph.D.
Department of Agricultural Engineering
University of Kentucky
Lexington, Kentucky

D. B. Brooker, Ph.D.
Professor Emeritus
Department of Agricultural Engineering
University of Missouri
Columbia, Missouri

R. H. Brown, P.E., Ph.D.
Chairman Emeritus
Division of Agricultural Engineering
University of Georgia
Athens, Georgia

R. R. Bruce, Ph.D.
Soil Scientist
Southern Piedmont Research Center
USDA
Watkinsville, Georgia

J. L. Butler, Ph.D.
Research Leader
Crop Systems Research Unit
USDA
Tifton, Georgia

W. J. Chancellor, Ph.D.
Professor
Department of Agricultural Engineering
University of California
Davis, California

J. L. Chesness, Ph.D.
Professor
Department of Agricultural Engineering
University of Georgia
Athens, Georgia

C. M. Christensen, Ph.D.
Professor Emeritus
Department of Plant Pathology
University of Minnesota
St. Paul, Minnesota

D. S. Chung, Ph.D.
Professor
Department of Agricultural Engineering
Kansas State University
Manhattan, Kansas

C. J. W. Drablos, Ph.D.
Professor
Department of Agricultural Engineering
University of Illinois
Urbana, Illinois

W. C. Fairbank
Extension Agricultural Engineer
Department of Soil and Environmental
Science
University of California
Riverside, California

P. R. Goodrich, Ph.D.
Associate Professor
Department of Agricultural Engineering
University of Minnesota
St. Paul, Minnesota

J. W. Goodrum, Ph.D
Associate Professor
Department of Agricultural Engineering
University of Georgia
Athens, Georgia

W. C. Hammond, Ph.D.
Head
Department of Extension Engineering
Cooperative Extension Services
Athens, Georgia

P. K. Harein, Ph.D.
Department of Entomology, Fisheries, &
Wildlife
University of Minnesota
St. Paul, Minnesota

J. C. Hayes, Ph.D.
Department of Agricultural Engineering
Clemson University
Clemson, South Carolina

J. G. Hendrick
Agricultural Engineer
National Tillage Machinery Lab
USDA
Auburn, Alabama

T. A. Howell, Ph.D.
Agricultural Engineer
Conservation and Production Research Lab
USDA
Bushland, Texas

R. W. Irwin, Ph.D.
Professor
School of Engineering
University of Guelph
Guelph, Ontario, Canada

F. C. Ives, P.E.
Design Engineer
Soil Conservation Service
USDA
Champaign, Illinois

J. M. Laflen, Ph.D.
Research Leader
National Soil Erosion Research Lab
USDA
West Lafayette, Indiana

J. H. Lehr, Ph.D.
Executive Director
National Water Well Association
Dublin, Ohio

W. D. Lembke, Ph.D.
Professor Emeritus
Department of Agricultural Engineering
University of Illinois
Urbana, Illinois

L. Lyles, Ph.D.
Research Leader
Wind Erosion Unit
USDA
Manhattan, Kansas

H. B. Manbeck, Ph.D.
Professor
Department of Agricultural Engineering
Pennsylvania State University
Univerisity Park, Pennsylvania

J. R. Miner, Ph.D.
Associate Director and Professor
Department of International Research and
Development
Oregon State University
Corvallis, Oregon

C. H. Moss, P.E.
Manager Civil Engineer
T. E. Stivers Organization, Inc.
Decatur, Georgia

B. H. Nolte, Ph. D.
Professor
Department of Agricultural Engineering
Ohio State University
Columbus, Ohio

J. C. Nye
Professor
Department of Agricultural Engineering
Louisiana State University
Baton Rouge, Louisiana

C. H. Pair
Engineer
USDA
Boise, Idaho

L. K. Pickett, Ph.D.
Senior Project Engineer
Department of Advanced Engineering
Case International
Hinsdale, Illinois

J. H. Poehlman
Largo, Florida

L. M. Safley, Jr., Ph.D.
Assistant Professor
Department of Agricultural Engineering
University of Tennessee
Knoxville, Tennessee

G. O. Schwab, Ph.D.
Professor Emeritus
Department of Agricultural Engineering
Ohio State University
Columbus, Ohio

Hollis Shull
Engineer
USDA
University of Nebraska
Lincoln, Nebraska

J. W. Simons
Research Associate
Department of Agricultural Engineering
University of Georgia
Athens, Georgia

R. P. Singh, Ph.D.
Professor
Department of Agricultural Engineering
University of California
Davis, California

R. E. Sneed, Ph.D.
Professor
Department of Biological and Agricultural
Engineering
North Carolina State University
Raleigh, North Carolina

J. M. Steichen, Ph.D.
Associate Professor
Department of Agricultural Engineering
Kansas State University
Manhattan, Kansas

C. W. Suggs, Ph.D.
Professor
Department of Biological and Agricultural
Engineering
North Carolina State University
Raleigh, North Carolina

J. M. Sweeten, Ph.D.
Extension Agricultural Engineer
Texas Agricultural Extension Service
Texas A&M University
College Station, Texas

J. R. Talbot
National Soil Engineer
USDA
Washington, D. C.

E. D. Threadgill, Ph.D.
Chairman
Department of Agricultural Engineering
University of Georgia
Athens, Georgia

D. H. Vanderholm
Associate Dean
Institute of Agriculture and Natural Resources
University of Nebraska
Lincoln, Nebraska

G. L. Van Wicklen, Ph.D.
Assistant Professor
Department of Agricultural Engineering
University of Georgia
Athens, Georgia

N. L. West
Project Engineer
John Deere Harvester
East Moline, Illinois

I. L. Winsett
Sales Engineer
Ronk Electrical Indusrties
Nokomis, Illinois

TABLE OF CONTENTS

Volume II

EROSION CONTROL ENGINEERING

DRAINAGE ENGINEERING

Volume III

PHYSIOLOGICAL PARAMETERS AND REQUIREMENTS OF LIVESTOCK POULTRY

STRUCTURAL DESIGNS, REQUIREMENTS, AND SYSTEMS

ELECTRICAL SYSTEMS AND APPLIANCES FOR AGRICULTURAL STRUCTURES

FEED AND CROP STORAGES

APPENDIX

Erosion Control Engineering

WATER EROSION CONTROL

John M. Laflen

INTRODUCTION

Soil erosion on agricultural lands is an international problem. In the U.S., soil erosion rates currently are a threat to the ability to produce agricultural goods to supply national and international demands. Troeh et al.[1] have detailed the historical effects of soil erosion throughout the world.

Water erosion is caused by rainfall, runoff detachment, and transport. In this chapter a method for predicting water erosion for the ways that man might manage, use, or treat his land will be presented. The necessary equations and coefficients will be given. Then, among the many alternatives, the use, management, or treatment of a specific land resource can be selected so that water erosion can be controlled at the desired level.

No attempt will be made to present the technology for estimating sediment yields from watersheds or for estimating gully erosion.

UNIVERSAL SOIL LOSS EQUATION

The major factors affecting soil erosion are the climate, soil, topography, land use, and practices for erosion control. During the 1930s when erosion first became a major national issue, plots to measure soil erosion, and to evaluate means to control it, were established in several areas of the U.S. Based on these data, equations were developed for prediction of soil erosion that were applicable within a region. In 1958, Wischmeier and Smith[2] isolated an important rainfall factor that explained much of the storm-to-storm variance in soil erosion on fallow plots at a single location. The same rainfall factor was found to be an excellent variable for explaining soil loss at all the locations tested. Then, using this rainfall variable, the effect of certain land use, topography, soil, or erosion treatment could be compared across regions. This allowed the development of a system for predicting soil erosion, commonly called the Universal Soil Loss Equation (USLE):[3]

$$A = R \ K \ LS \ C \ P \tag{1}$$

where A is computed soil loss for the time period for R (t/ha), R is a rainfall factor, K is a soil erodibility factor, LS is a length slope factor, C is a cover-management factor, and P is a support practice factor.

The dimensions of the various factors, as used herein,[4] are

A	metric tons/hectare or t/ha
R	$\dfrac{\text{megajoules·millimeter}}{\text{hectare·mm·hr}}$ or $\dfrac{\text{MJ·mm}}{\text{ha·mm·hr}}$
K	$\dfrac{\text{metric tons·hectare·hour}}{\text{hectare·megajoules·millimeter}}$ or $\dfrac{\text{t·ha·hr}}{\text{ha·MJ·mm}}$
LS, C, P	dimensionless

RAINFALL FACTOR

The rainfall factor, for qualifying storms, is the product of the rainfall energy for a storm and the maximum 30-min rainfall intensity within a storm. Storms having less than 13 mm

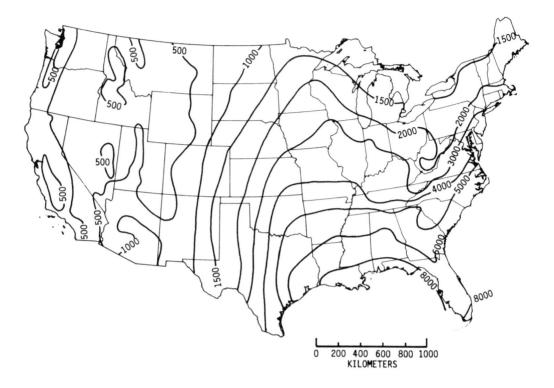

FIGURE 1.   Annual rainfall factor values for the U.S. (Redrawn from Wischmeier and Smith.[3])

of rainfall, and separated from other rain periods by more than 6 hr, are not qualifying storms, unless they have at least 6 mm of rain in a 15-min period. The rainfall factor can be summed over all qualifying storms within a year, or some other time period, to determine rainfall factor values for a desired time period. Rainfall kinetic energy (KE) per millimeter of rainfall is given by Equations 2 or 3 as

$$KE = 0.119 + 0.087 \log_{10}(I), \quad I \leq 76 \text{ mm/hr} \tag{2}$$

$$KE = 0.283 \qquad\qquad\qquad I > 76 \text{ mm/hr} \tag{3}$$

where I is rainfall intensity (mm/hr). The storm value of the rainfall factor (R_s) is

$$R_s = I_m \sum_{i=1}^{n} (KE_i) P_i \tag{4}$$

where I_m is the maximum 30-min rainfall intensity for a storm, KE_i is the kinetic energy in the ith interval of the n intervals for a storm, and P_i is precipitation amount within the ith interval. Average annual R values for the U.S. are shown in Figure 1.

The science of estimating soil erosion from snowmelt and thaw is still emerging. Wischmeier and Smith[3] provide a means of making an estimate of the rainfall factor values for thaw and snowmelt and should be consulted in those cases. Water erosion under these conditions is a topic of current research interest, and results may be available in the near future.

SOIL ERODIBILITY FACTOR

The soil erodibility factor (K) is the rate of soil loss per unit rainfall factor as measured on a plot 22.1 m long of 9% slope maintained in continuous fallow.[5] The K value for a soil, where the silt fraction is less than 70%, is given by[6]

$$K = 0.28 \times 10^{-6}(S_iC)^{1.14}(12 - OM) + 0.0043(S - 2) + 0.0033(P - 3) \quad (5)$$

where S_i is the percent of soil material having diameters greater than 0.002 mm and less than 0.10 mm, C is the percent of soil material greater than 0.002 mm in diameter, OM is the percent organic matter, S is a structure code (1 = very fine granular, 2 = fine granular, 3 = medium or coarse granular, and 4 = blocky, platy, or massive structure), and P is a permeability code (1 = rapid, 2 = moderate to rapid, 3 = moderate, 4 = slow to moderate, 5 = slow, and 6 = very slow). Wischmeier et al.[6] provide a nomograph for determination of K that includes silt fractions greater than 70%.

LENGTH-SLOPE FACTOR

The length-slope (LS) factor for a uniform slope is given by

$$LS = \left(\frac{\lambda}{22.1}\right)^m (0.065 + 0.046\ S + 0.0065\ S^2) \quad (6)$$

where λ is slope length (in meters), S is slope (percent), and m has values of 0.5 for S > 5, 0.4 for 3.5 < S < 4.5, 0.3 for 1 < S < 3, and 0.2 for S < 1. Data used to determine the relationship in Equation 6 were from studies where slopes were between 3 and 18% and lengths from 9 to 91 m. Slope length is the distance from the point where overland flow originates to the point where runoff enters a well-defined channel or the slope flattens enough for deposition to begin.

The LS factor for nonuniform slopes can be computed. First, the slope length is divided into N equal length segments, each with a nearly uniform slope. The fraction of the total soil loss from the entire slope occurring on that increment (if the entire slope were uniform) would be

$$\text{Soil loss} = \frac{I^{m+1} - (I - 1)^{m+1}}{N^{m+1}} \quad (7)$$

where I is the segment sequence number, with numbering beginning at the upper end of the slope, and m is as above. Then, for each segment of the nonuniform slope, compute an LS factor for that segment using its slope and the length of the entire slope. Next, multiply the LS value for each segment by the fraction from that segment (Equation 7), and sum these over all segments to compute the LS value for the entire slope. LS for the nonuniform slope broken into N uniform length segments with a constant slope (S_i) within each segment is given by

$$LS = \frac{1}{N^{m+1}} \left(\frac{\lambda}{22.1}\right) \sum_{I=1}^{N} [I^{m+1} - (I - 1)^{m+1}][0.065 + 0.046\ S_i + 0.0065\ s_i^2] \quad (8)$$

where λ is the entire slope length and m is as before. For more complex situations, where soil erodibility changes along a slope, or segments are of unequal length, Wischmeier and Smith[3] should be consulted.

COVER-MANAGEMENT FACTOR

The cover-management factor (C) is the ratio of soil loss for a specific condition to soil loss from a continuous fallow.[7] It expresses the combined effect of a large number of factors. These factors include canopy, tillage, crop residue, roughness, antecedent moisture, buried residue, and prior land use (cropping history).

C-values have been determined for many conditions from side-by-side comparisons of soil erosion under specific conditions to soil erosion under the same soil, climate, and topographic conditions, but with the surface maintained vegetation free and with regular cultivation so that the surface is maintained similarly to that of a seedbed prepared for planting. Many such comparisons have been made at many locations across the U.S.

C-values have also been estimated, where data are limited, as the product of a series of subfactors. These subfactors usually express the effect of residue cover, crop canopy, incorporated residues, tillage, and prior land use. Generally, C-values for cropped conditions have come from side-by-side comparisons, while for most other situations, the subfactor approach has been used.

C-values in this chapter are expressed in percentages. They must be divided by 100 for use in Equation 1.

C for Cropping

Since the effects of the factors above on soil erosion vary widely during the year for a particular crop, the cropping season has been separated into six distinct crop stages, each having a C-value that expresses the average ratio of soil erosion during that period to soil erosion for the fallow condition. The stages are

F   Inversion plowing to secondary tillage
SB   Secondary tillage to 10% crop canopy
1   10% crop canopy to 50% crop canopy
2   50% crop canopy to 75% crop canopy
3   75% crop canopy to harvest
4   Harvest to plowing or new seeding

C-values are given in Tables 1 to 5 for various crops, crop stages, tillage systems, and crop rotations. Use of the tables requires considerable judgment. The footnotes should be examined in detail.

Lines in Table 1 are frequently delineated on the basis of spring residue weight (kg/ha) and the percentage of the ground surface covered by crop residue after planting. The spring residue weight is the weight of residue in the spring before spring tillage.

C for Construction Sites

The USLE is also used to estimate soil erosion from construction sites. C-values for no mulch and for mulches of straw, hay, wood chips, and crushed stone are given in Table 6. Soil loss ratios for many conditions on construction slopes can be computed using C-values from cropland, if good judgment is used in selecting cropland values for comparable surface and cropping condition on construction sites, and can be varied over time, similar to C for cropped conditions, when conditions warrant. Meyer and Ports[8] have performed considerable work in adapting the USLE to construction sites.

C for Forests

Dissmeyer and Foster[9,10] have determined C-values for forest conditions, based on the subfactor approach. For best estimates, the reader is referred to their publications. C-values for undisturbed forest land are given in Table 7 and for mechanically prepared woodland sites in Table 8.

Table 1
RATIO OF SOIL LOSS FROM CROPLAND TO CORRESPONDING LOSS FROM CONTINUOUS FALLOW

Line no.	Cover, crop sequence, and management[a]	Spring residue[b] (kg/ha)	Cover after plant[c] (%)	Soil loss ratio[d] for crop stage period and canopy cover[e] (%)							
				F	SB	1	2	3:80	90	96	4L[f]
	Corn after C, GS, G, or COT in meadowless systems Moldboard plow, conv till										
1	Rdl, sprg TP	5000	—	31	55	48	38	—	—	20	23
2		3800	—	36	60	52	41	—	24	20	30
3		3000	—	43	64	56	43	32	25	21	37
4		2250	—	51	68	60	45	33	26	22	47
5	RdL, fall TP	HP[b]	—	44	65	53	38	—	—	20	—
6		GP	—	49	70	57	41	—	24	20	—
7		FP	—	57	74	61	43	32	25	21	—
8		LP	—	65	78	65	45	32	26	22	—
9	RdR, sprg TP	HP	—	66	74	65	47	—	—	22	56[g]
10		GP	—	67	75	66	47	—	27	23	62
11		FP	—	68	76	67	48	35	27	—	69
12		LP	—	69	77	68	49	35	—	—	74
13	RdR, fall TP	HP	—	76	82	70	49	—	—	22	—
14		GP	—	77	83	71	50	—	27	23	—
15		FP	—	78	85	72	51	35	27	—	—
16		LP	—	79	86	73	52	35	—	—	—
17	Wheeltrack pl, RdL, TP[h]	5000	—	—	31	27	25	—	—	18	23
18		3800	—	—	36	32	30	—	22	18	30
19		3000	—	—	43	36	32	29	23	19	37
20		2250	—	—	51	43	36	31	24	20	47
21	Deep off-set disk or disk plow	5000	10	—	45	38	34	—	—	20	23
22		3800	10	—	52	43	37	—	24	20	30
23		3000	5	—	57	48	40	32	25	21	37
24		2250	—	—	61	51	42	33	26	22	47
25	No-till plant in crop residue[i]	6700	95	—	2	2	2	—	—	2	14
26		6700	90	—	3	3	3	—	—	3	14
27		5000	80	—	5	5	5	—	—	5	15
28		3800	70	—	8	8	8	—	8	6	19
29		3800	60	—	12	12	12	12	9	8	23
30		3800	50	—	15	15	14	14	11	9	27
31		3000	40	—	21	20	18	17	13	11	30
32		3000	30	—	26	24	22	21	17	14	36
	Chisel, shallow disk, or fld cult, as only tillage										
33	On moderate slopes	6700	70	—	8	8	7	—	—	7	17
34			60	—	10	9	8	—	—	8	17
35			50	—	13	11	10	—	—	9	18
36			40	—	15	13	11	—	—	10	19
37			30	—	18	15	13	—	—	12	20
38			20	—	23	20	18	—	—	16	21
39	Do.	5000	70	—	9	8	7	—	—	7	18
40			60	—	12	10	9	—	—	8	18
41			50	—	14	13	11	—	—	9	19
42			40	—	17	15	13	—	—	10	20
43			30	—	21	18	15	—	—	13	21
44			20	—	25	22	19	—	—	16	22
45	Do.	3800	60	—	13	11	10	—	10	8	20
46			50	—	16	13	12	—	12	9	24
47			40	—	19	17	16	—	14	11	25

Table 1 (continued)
RATIO OF SOIL LOSS FROM CROPLAND TO CORRESPONDING LOSS FROM CONTINUOUS FALLOW

Line no.	Cover, crop sequence, and management[a]	Spring residue[b] (kg/ha)	Cover after plant[c] (%)	Soil loss ratio[d] for crop stage period and canopy cover[e] (%)							
				F	SB	1	2	3:80	90	96	4L[f]
48			30	—	23	21	19	—	17	14	26
49			20	—	29	25	23	—	21	16	27
50			10	—	36	32	29	—	24	20	30
51	Do.	3000	50	—	17	16	15	15	13	10	29
52			40	—	21	20	19	19	15	12	30
53			30	—	25	23	22	22	18	14	32
54			20	—	32	29	28	27	22	17	34
55			10	—	41	36	34	32	25	21	37
56	Do.	2200	40	—	23	21	20	20	15	12	37
57			30	—	27	25	24	23	19	15	39
58			20	—	35	32	30	28	22	18	42
59			10	—	46	42	38	33	26	22	47
	On slopes >12%										
60	Lines 33—59 times factor of	—	—	—	1.3	1.3	1.1	1.0	1.0	1.0	1.0
	Disk or harrow after spring chisel or fld cult										
	Lines 33—59 times factor of										
61	On moderate slopes	—	—	—	1.1	1.1	1.1	1.0	1.0	1.0	1.0
62	On slopes >12%	—	—	—	1.4	1.4	1.2	1.0	1.0	1.0	1.0
	Ridge plants[j]										
	Lines 33—59 times factor of										
63	Rows on contour[k]	—	—	—	0.7	0.7	0.7	0.7	0.7	0.7	0.7
64	Rows U/D slope <12%	—	—	—	0.7	0.7	1.0	1.0	1.0	1.0	1.0
65	Rows U/D slope >12%	—	—	—	0.9	0.9	1.0	1.0	1.0	1.0	1.0
	Till plant										
	Lines 33—59 times factor of										
66	Rows on contour[k]	—	—	—	0.7	0.85	1.0	1.0	1.0	1.0	1.0
67	Rows U/D slope <7%	—	—	—	1.0	1.0	1.0	1.0	1.0	1.0	1.0
	Strip till one fourth of row spacing										
68	Rows on contour[k]	5000	60[i]	—	12	10	9	—	—	8	23
69		3800	50	—	16	14	12	—	11	10	27
70		3000	40	—	22	19	17	17	14	12	30
71		2200	30	—	27	23	21	20	16	13	36
72	Rows U/D slope	5000	60[i]	—	16	13	11	—	—	9	23
73		3800	50	—	20	17	14	—	12	11	27
74		3000	40	—	26	22	19	17	14	12	30
75		2200	30	—	31	26	23	20	16	13	36
	Vari-till										
76	Rows on contour[k]	3800	40	—	13	12	11	—	—	11	22
77		3800	30	—	16	15	14	14	13	12	26
78		3000	20	—	21	19	19	19	16	14	34
	Corn after WC of ryegrass or wheat seeded in C stubble										
	WC reaches stemming stage										
79	No-till p1 in killed WC	4500	—	—	7	7	7	—	7	6	m
80		3400	—	—	11	11	11	11	9	7	
81		2200	—	—	15	15	14	14	11	9	
82		1700	—	—	20	19	18	18	14	11	
	Strip till one fourth row space										
83	Rows U/D slope	4500	—	—	13	12	11	—	11	9	m

Table 1 (continued)
RATIO OF SOIL LOSS FROM CROPLAND TO CORRESPONDING LOSS FROM CONTINUOUS FALLOW

Line no.	Cover, crop sequence, and management[a]	Spring residue[b] (kg/ha)	Cover after plant[c] (%)	Soil loss ratio[d] for crop stage period and canopy cover[e] (%)							
				F	SB	1	2	3:80	90	96	4L[f]
84		3400	—	—	18	17	16	16	13	10	
85		2200	—	—	23	22	20	19	15	12	
86		1700	—	—	28	26	24	22	17	14	
87	Rows on contour[k]	4500	—	—	10	10	10	—	10	8	m
88		3400	—	—	15	15	15	15	12	9	
89		2200	—	—	20	20	19	19	15	12	
90		1700	—	—	25	24	23	22	17	14	
91	TP, conv seedbed	4500	—	36	60	52	41	—	24	20	m
92		3400	—	43	64	56	43	31	25	21	
93		2200	—	51	68	60	45	33	26	22	
94		1700	—	61	73	64	47	35	27	23	
	WC succulent blades only										
95	No-till p1 in killed WC	3400	—	—	11	11	17	23	18	16	m
96		2200	—	—	15	15	20	25	20	17	
97		1700	—	—	20	20	23	26	21	18	
98		1100	—	—	26	26	27	27	22	19	
99	Strip till one fourth row space	3400	—	—	18	18	21	25	20	17	m
100		2200	—	—	23	23	25	27	21	18	
101		1700	—	—	28	28	28	28	22	19	
102		1100	—	—	33	33	31	29	23	20	
	Corn in sod-based systems										
	No-till p1 in killed sod										
103	<6 t/ha hay yld	—	—	—	1	1	1	—	1	1	1
104	<5 t/ha hay yld	—	—	—	2	2	2	2	2	2	2
	Strip till, >6 t/ha M										
105	50% cover, tilled strips	—	—	—	2	2	2	—	2	2	4
106	20% cover, tilled strips	—	—	—	3	3	3	—	3	3	5
	Strip till, <5 t/ha M										
107	40% cover, tilled strips	—	—	—	4	4	4	4	4	4	6
108	20% cover, tilled strips	—	—	—	5	5	5	5	5	5	7
	Other tillage after sod		n	n	n	n	n	n	n	n	n
	Corn after soybeans										
109	Sprg TP, conv till	HP	—	40	72	60	48	—	—	25	29
110		GP	—	47	78	65	51	—	30	25	37
111		FP	—	56	83	70	54	40	31	26	44
112	Fall TP, conv till	HP	—	47	75	60	48	—	—	25	—
113		GP	—	53	81	65	51	—	30	25	—
114		FP	—	62	86	70	54	40	31	26	—
115	Fall and sprg chisel or cult	HP	30°	—	40	35	29	—	—	23	29
116		GP	25	—	45	39	33	—	27	23	37
117		GP	20	—	51	44	39	34	27	23	37
118		FP	15	—	58	51	44	36	28	23	44
119		LP	10	—	67	59	48	36	28	23	54
120	No-till p1 in crop res'd	HP	40°	—	25	20	19	—	14	11	26
121		GP	30	—	33	29	25	22	18	14	33
122		FP	20	—	44	38	32	27	23	18	40
	Beans after corn										
123	Sprg TP, RdL, conv. till	HP	—	33	60	52	38	—	20	17	p
124		GP	—	39	64	56	41	—	21	18	
125		FP	—	45	68	60	43	29	22	—	

Table 1 (continued)
RATIO OF SOIL LOSS FROM CROPLAND TO CORRESPONDING LOSS FROM CONTINUOUS FALLOW

Line no.	Cover, crop sequence, and management[a]	Spring residue[b] (kg/ha)	Cover after plant[c] (%)	Soil loss ratio[d] for crop stage period and canopy cover[e] (%)							
				F	SB	1	2	3:80	90	96	4L[f]
126	Fall TP, RdL, conv. till	HP	—	45	69	57	38	—	20	17	p
127		GP	—	52	73	61	41	—	21	18	
128		FP	—	59	77	65	43	29	22	—	
	Chisel or fld cult			q	q	q	q	q	q	q	p
	Beans after beans			r	r	r	r	r	r	r	p
	Grain after C, G, GS, COT[s]										
129	In disked residues	5000	70	—	12	12	11	7	4	2	t
130		3800	60	—	16	14	12	7	4	2	
131			50	—	22	18	14	8	5	3	
132			40	—	27	21	16	9	5	3	
133			30	—	32	25	18	9	6	3	
134			20	—	38	30	21	10	6	3	
135	Do.	3000	40	—	29	24	19	9	6	3	t
136			20	—	43	34	24	11	7	4	
137			10	—	52	39	27	12	7	4	
138	Do.	2200	30	—	38	30	23	11	7	4	t
139			20	—	46	36	26	12	7	4	
140			10	—	56	43	30	13	8	5	
141	In disked stubble, RdR	—	—	—	79	62	42	17	11	6	t
142	Winter G after fall TP, RdL	HP	—	31	55	48	31	12	7	5	t
143		GP	—	36	60	52	33	13	8	5	
144		FP	—	43	64	56	36	14	9	5	
145		LP	—	53	68	60	38	15	10	6	
	Grain after summer fallow										
146	With grain residues	200	10	—	70	55	43	18	13	11	u
147		550	30	—	43	34	23	13	10	8	
148		850	40	—	34	27	18	10	7	7	
149		1100	50	—	26	21	15	8	7	6	
150		1700	60	—	20	16	12	7	5	5	
151		2200	70	—	14	11	9	7	5	5	
152	With row crop residues	350	5	—	82	65	44	19	14	12	u
153		550	15	—	62	49	35	17	13	11	
154		850	23	—	50	40	29	14	11	9	
155		1100	30	—	40	31	24	13	10	8	
156		1700	45	—	31	24	18	10	8	7	
157		2200	55	—	23	19	14	8	7	5	
158		2800	65	—	17	14	12	7	5	4	
	Potatoes										
159	Rows with slope	—	—	43	64	56	36	26	19	16	
160	Contoured rows, ridged when canopy cover is about 10%[k]	—	—	43	64	28	18	13	10	8	

[a]   Symbols: B, soybeans; C, corn; conv till, plow, disk, and harrow for seedbed; cot, cotton; F, rough fallow; fld cult, field cultivator; G, small grain; GS, grain sorghum; M, grass and legume meadow, at least 1 full year; pl, plant; RdL, crop residues left on field; RdR, crop residues removed; SB, seedbed period; sprg, spring; TP, plowed with moldboard; WC, winter cover crop; — , insignificant or an unlikely combination of variables.

[b]   Dry weight after winter loss and reduction by grazing or partial removal: 5000 kg/ha represents 2.5 to 3 t/ha corn; 3800 kg/ha, 2 to 2.5 t/ha; 3000 kg/ha, 1.5 to 2 t/ha; 2250 kg/ha, 1 to 1.5 t/ha; with normal 30% winter loss. For RdR or fall-plow practices, these four productivity levels are indicated by HP, GP, FP, and LP, respectively (high, good, fair, and low productivity). In lines 79 to 102, this column indicates dry weight of the winter cover crop.

Table 1 (continued)
RATIO OF SOIL LOSS FROM CROPLAND TO CORRESPONDING LOSS FROM CONTINUOUS FALLOW

^c Percentage of soil surface covered by plant residue mulch after crop seeding. The difference between spring residue and that on the surface after crop seeding is reflected in the soil loss ratios as residues mixed with the topsoil.

^d The soil loss ratios, given as percentages, assume that the indicated crop sequence and practices are followed consistently. One-year deviations from normal practices do not have the effect of a permanent change. Linear interpolation between lines is recommended when justified by field conditions.

^e Crop stage periods are as defined earlier. The three columns for crop stage 3 are for 80, 90, and 96 to 100% canopy cover at maturity.

^f Column 4L is for all residues left on field. Corn stalks partially standing as left by some mechanical pickers. If stalks are shredded and spread by picker, select ratio from Table 4. When residues are reduced by grazing, take ratio from lower spring-residue line.

^g Period 4 values in lines 9 to 12 are for corn stubble (stover removed).

^h Inversion plowed, no secondary tillage. For this practice, residues must be left and incorporated.

ⁱ Soil surface and chopped residue of matured preceding crop undisturbed except in narrow slots in which seeds are planted.

^j Top of old row ridge sliced off, throwing residues and some soil into furrow areas. Reridging assumed to occur near end of crop stage 1.

^k Where lower soil loss ratios are listed for rows on the contour, this reduction is in addition to the standard field contouring credit. The P-value for contouring is used with these reduced loss ratios.

^l Field-average percent cover; probably about three fourths of percent cover on undisturbed strips.

^m Divide the winter-cover period into crop stages for the seeded cover and use lines 132 to 145.

ⁿ Select the appropriate line for the crop, tillage, and productivity level and multiply the listed soil loss ratios by sod residual factors from Table 5.

^o Spring residue may include carryover from prior corn crop.

^p See Table 4.

^q Use values from lines 33 to 62 with appropriate dates and lengths of crop stage periods for beans in the locality.

^r Values in lines 109 to 122 are best available estimates, but planting dates and lengths of crop stages may differ.

^s When meadow is seeded with the grain, its effect will be reflected through higher percentages of cover in crop stages 3 and 4.

^t Ratio depends on percent cover. See Table 4.

^u See Item 12, Table 3.

Adapted from Wischmeier and Smith.[3]

C for Pasture, Range, and Idle Land

C-values for pasture, range, and idle land are given in Table 9.[11] These were derived by evaluating three factors having a major influence on C-values: (1) crop residue — crop material in contact with the ground surface, (2) crop canopy, and (3) residual land use and tillage effects. Data to evaluate C-values for pasture, range, and idle land are scarce, hence the use of the subfactor approach.

SUPPORT-PRACTICE FACTOR

The support-practice factor (P) is the ratio of soil loss with a specific practice to soil loss with up- and downhill farming. Support practices include contouring, contour strip cropping, and terracing, on cropland.

P-values for contouring are given in Table 10 by land slope. Also given in Table 10 are maximum slope lengths; when slope length exceeds these limits, P-values less than one are not advisable.

P-values for contour strip cropping are given in Table 11. As for contouring, there are limits given, both for slope length and strip width. Additionally, the P-values vary with crop rotation. Minor deviations from strip widths can be tolerated to accommodate farm machinery of different widths.

Table 2
C FACTORS FOR COTTON

		65	80	95
Expected final canopy percent cover:		65	80	95
Estimated initial percent cover from defoliation + stalks down:		30	45	60

Practice no.	Tillage operation(s)[a]	Soil loss ratio[b] (%)		

Cotton Annually

Practice no.	Tillage operation(s)[a]			
1	None			
	Defoliation to Dec. 31	36	24	15
	Jan. 1 to Feb. or Mar. tillage			
	Cot Rd only	52	41	32
	Rd and 20% cover vol veg[b]	32	26	20
	Rd and 30% cover vol veg	26	20	14
2	Chisel plow soon after cot harvest			
	Chiseling to Dec. 31	40	31	24
	Jan. 1 to spring tillage	56	47	40
3	Fall disk after chisel			
	Disking to Dec. 31	53	45	37
	Jan. 1 to spring tillage	62	54	47
4	Chisel plow Feb.—Mar., no prior tillage			
	Cot Rd only	50	42	35
	Rd and 20% vol veg	39	33	28
	Rd and 30% vol veg	34	29	25
5	Bed ("hip") Feb.—Mar., no prior tillage			
	Cot Rd only	100	84	70
	Rd and 20% vol veg	78	66	56
	Rd and 30% vol veg	68	58	50
	Split ridges and plant after hip, or disk and plant after chisel (SB)			
	Cot Rd only	61	54	47
	Rd and 20% vol veg	53	47	41
	Rd and 30% vol veg	50	44	38
	Crop stage 1			
	Cot Rd only	57	50	43
	Rd and 20% vol veg	49	43	38
	Rd and 30% vol veg	46	41	36
	Crop stage 2	45	39	34
	Crop stage 3	40	27	17
6	Bed (hip) after 1 prior tillage			
	Cot Rd only	110	96	84
	Rd and 20% veg	94	82	72
	Rd and 30% veg	90	78	68
	Split ridges after hip (SB)			
	Cot Rd only	66	61	52
	Rd and 20 to 30% veg	61	55	49
	Crop stage 1			
	Cot Rd only	60	56	49
	Rd and 20 to 30% veg	56	51	46
	Crop stage 2	47	44	38
	Crop stage 3	42	30	19
7	Hip after two prior tillages			
	Cot Rd only	116	108	98
	Rd and 20—30% veg	108	98	88
	Split ridges after hip (SB)	67	62	57
8	Hip after three or more tillages	120	110	102
	Split ridges after hip (SB)	68	64	59

Table 2 (continued)
C FACTORS FOR COTTON

Expected final canopy percent cover:		65	80	95
Estimated initial percent cover from defoliation + stalks down:		30	45	60

Practice no.	Tillage operation(s)[a]	Soil loss ratio[b] (%)		
9	Conventional moldboard plow and disk			
	Fallow period	42	39	36
	Seedbed period	68	64	59
	Crop stage 1	63	59	55
	Crop stage 2	49	46	43
	Crop stage 3	44	32	22
	Crop stage 4 (see practices 1, 2, and 3)			

Cotton After Sod Crop

For the first or second crop after a grass or grass-and-legume meadow has been turnplowed, multiply values given in the last five lines above by sod residual factors from Table 5.

Cotton After Soybeans

Select values from above and multiply by 1.25.

[a] Rd, crop residue; vol veg, volunteer vegetation.

[b] Alternate procedure for estimating the soil loss ratios. The ratios given above for cotton are based on estimates for reductions in percent cover through normal winter loss and by the successive tillage operations. Where the reductions in percent cover by winter loss and tillage operations are small, the following procedure may be used to compute soil loss ratios for the preplant and seedbed periods. Multiply the factor selected below by $e^{-0.025\,RC}$, where RC is percent of the land surface covered by crop residue to determine soil loss ratio. The factor below expresses the effect of land-use residual, surface roughness, and porosity.

Productivity level	No tillage	Rough surface	Smoothed surface
High	0.66	0.50	0.56
Medium	0.71	0.54	0.61
Poor	0.75	0.58	0.65

Values for the bedded period on slopes of less than 1% should be estimated at twice the value computed above for rough surfaces.

Adapted from Wischmeier and Smith.[3]

P-values for contour farmed terraced fields are given in Table 12. Note that values here do not include the reduced slope length, which must be included in LS computations. Two values are given for farm planning. One is to estimate the soil erosion between terraces and the second is to estimate the soil erosion between terraces and to credit the terrace in reducing soil loss from the field. Generally, the strip crop factor should be used.

Table 3
C-FACTORS FOR CONDITIONS NOT EVALUATED IN TABLE 1

Cotton
  See Table 2
Crop stage 4 for rowcrops
  Stalks broken and partially standing: use column 4L
  Stalks standing after hand picking: column 4L times 1.15
  Stalks shredded without soil tillage: see Table 4
  Fall chisel: select values from lines 33—62, seedbed column.
Crop stage 4 for small grain
  See Table 4
Double cropping
  Derive annual C-value by selecting from Table 1 the soil loss percentages for the successive crop stage periods
    of each crop
Established meadow, full-year percentages

Grass and legume mix	>6 ton/ha	0.4
Do.	4—6 ton/ha	0.6
Do.	<3 ton/ha	1.0
Sericea, after second year		1.0
Red clover		1.5
Alfalfa, lespedeza, and second-year sericea		2.0
Sweet clover		2.5

Meadow seeding without nurse crop
  Determine appropriate lengths of crop stage periods SB, 1, and 2 and apply values given for small grain seeding
Peanuts
  Comparison with soybeans is suggested
Pineapples
  Direct data not available; tentative values derived analytically are available from the SCS in Hawaii or the
    Western Technical Service Center at Portland, Oregon
Sorghum
  Select values given for corn, on the basis of expected crop residues and canopy cover
Sugarbeets
  Direct data not available; probably most nearly comparable to potatoes, without the ridging credit
Sugarcane
  Tentative values available from sources given for pineapples
Summer fallow in low-rainfall areas, use grain or row crop residues
  The approximate soil loss percentage after each successive tillage operation may be obtained from the following
    tabulation by estimating the percent surface cover after that tillage and selecting the column for the appropriate
    amount of initial residue; the given values (in percent) credit benefits of the residue mulch, residues mixed
    with soil by tillage, and the crop system residual

Percent cover by mulch	Initial residue (kg/ha)			
	>4500	3400	2200	1700
90	4	—	—	—
80	8	8[a]	—	—
70	12	13	14[a]	—
60	16	17	18[a]	19[a]
50	20	22	24	25[a]
40	25	27	30	32
30	29	33	37	39
20	35	39	44	48
10	47	55	63	68

  Winter cover seeding in row crop stubble or residues
    Define crop stage periods based on the cover seeding
    date and apply values from lines 129 to 145
  [a] For grain residue only.

Adapted from Wischmeier and Smith.[3]

Table 4
SOIL LOSS RATIOS (%) FOR CROP STAGE 4
WHEN STALKS ARE CHOPPED AND
DISTRIBUTED WITHOUT SOIL TILLAGE

Mulch cover[a]	Corn or sorghum (%)		Soybeans (%)		Stubble[d] (%)
	Tilled seedbed[b]	No-till	Tilled seedbed[b]	No-till in corn rd[c]	
20	48	34	60	42	48
30	37	26	46	32	37
40	30	21	38	26	30
50	22	15	28	19	22
60	17	12	21	16	17
70	12	8	15	10	12
80	7	5	9	6	7
90	4	3	—	—	4
95	3	2	—	—	3

[a] Part of a field surface directly covered by pieces of residue mulch.
[b] This column applies for all systems other than no-till.
[c] Cover after bean harvest may include an appreciable number of stalks carried over from the prior corn crop.
[d] For grain with meadow seeding, include meadow growth in percent cover and limit grain period 4 to 2 months. Thereafter, classify as established meadow.

Adapted from Wischmeier and Smith.[3]

APPLICATION

In practice, Equation 1 is usually rewritten as

$$A = RKLSP \sum_{i=1}^{n} f_i C_i \qquad (9)$$

where the period of time over which soil loss prediction is desired is broken into n periods, with C-values determined for each period, and the fraction of annual R, f_i, occurring within each period. These values are given for various areas of the U.S. (Table 13). The areas to which particular lines in Table 13 apply are shown in Figure 2. Note that in this chapter, C-values are expressed in percent in all the tables and must be divided by 100 before computing soil erosion.

The latter terms in Equation 9 is

$$\sum_{i=1}^{n} f_i C_i \qquad (10)$$

$$\sum_{i=1}^{n} f_i = 1 \qquad (11)$$

can be used to compare annual C-values for different cropping, rotations, and tillage systems. Estimates of such C-values are given in Table 14 for some of the more common crop managements expected.

Table 5

FACTORS TO CREDIT RESIDUAL EFFECTS OF TURNED SOD[a]

Crops	Hay yield (t/ha)	Factor of crop stage period (%)				
		F	SB and 1	2	3	4
First year after mead						
Row crop or grain	>7	25	40	45	50	60
	4—7	30	45	50	55	65
	<4	35	50	55	60	70
Second year after mead						
Row crop	>7	70	80	85	90	95
	4—7	75	85	90	95	100
	<4	80	90	95	100	100
Spring grain	>7	—	75	80	85	95
	4—7	—	80	85	90	100
	<4	—	85	90	95	100
Winter grain	>7	—	60	70	85	95
	4—7	—	65	75	90	100
	<4	—	70	85	95	100

[a] These factors are to be multiplied by the appropriate soil loss percentages selected from Table 1. They are directly applicable for sod-forming meadows of at least 1 full year duration, plowed not more than 1 month before final seedbed preparation. When sod is fall plowed for spring planting, the listed values for all crop stage periods are increased by adding 0.02 for each additional month by which the plowing precedes spring seedbed preparation. For example, September plowing would precede May disking by 8 months and 0.02(8-1) or 0.14 would be added to each value in the table. For nonsod-forming meadows, such as sweetclover or lespedeza, multiply the factors by 1.2. When the computed value is greater than 1.0, use as 1.0.

Adapted from Wischmeier and Smith.[3]

When soil loss is to be limited by some value T, then values of C can be determined by

$$C = \frac{T}{RKLSP} \tag{12}$$

In these cases, having previously determined values of C for applicable cropping and conservation practices aids in selection of cropping and conservation strategies to limit soil erosion. Generally, such tables for a particular locality are available from the U.S. Department of Agriculture, Soil Conservation Service, usually at both state and county offices.

An example of use of the USLE is presented below. Assume the following conditions:

1. Location: Central Illinois
2. Crop: Corn-soybean rotation, high productivity
3. Tillage: Fall chisel plow, spring disk after corn; no-tillage after soybeans
4. Soil: 60% silt, 10% very fine sand, 10% clay, 3% organic matter, fine granular structure, moderate permeability
5. Topography: Slope — 7%, length 45 m
6. Farming practice: Up- and downhill

Then, R = 3200, K (Equation 5) = 0.050 LS (Equation 6) = 1.05, and values for f_i are

Table 6
C-VALUES AND LENGTH LIMITS FOR
CONSTRUCTION SLOPES

Type of mulch	Mulch rate (t/ha)	Land slope (%)	C (%)	Length limit[a] (m)
None	0	All	100	—
Straw or hay, tied down by anchoring and tacking equipment[b]	2	1—5	20	60
	2	6—10	20	30
	3	1—5	12	90
	3	6—10	12	45
	4	1—5	6	120
	4	6—10	6	60
	4	11—15	7	45
	4	16—20	11	30
	4	21—25	14	25
	4	26—33	17	15
	4	34—50	20	10
Crushed stone, 6 to 40 mm	300	<16	5	60
	300	16—20	5	45
	300	21—33	5	30
	300	34—50	5	25
	540	<21	2	90
	540	21—33	2	60
	540	34—50	2	45
Wood chips	15	<16	8	25
	15	16—20	8	15
	25	<16	5	45
	25	16—20	5	30
	25	21—33	5	25
	55	<16	2	60
	55	16—20	2	45
	55	21—33	2	30
	55	34—50	2	25

[a] Maximum slope length for which the specified mulch rate is considered effective. When this limit is exceeded, either a higher application rate or mechanical shortening of the effective slope length is required.

[b] When the straw or hay mulch is not anchored to the soil, C-values on moderate or steep slopes of soils having K values greater than 0.04 would be taken at double the values given in this table.

Adapted from Wischmeier and Smith.[3]

taken from line 16 in Table 9. The computations for an average annual C-value are given in Table 15. Then, from Equation 1, the average annual soil loss is

$$A = (3200)(0.050)(1.05)(0.133)(1) = 22.3 \text{ t/ha}$$

If maximum permissible average annual soil loss were about 11 t/ha, use of a P-factor of 0.5 would reduce soil loss to near the maximum permissible. For this slope and length, the P-value (Table 10) would be 0.5. An additional reduction could be achieved by further reducing the tillage operations after corn so that more residue would be on the surface in the spring.

Table 7
C-VALUES FOR UNDISTURBED FOREST LAND[a]

Percent of area covered by canopy of trees and undergrowth	Percent of area covered by duff at least 50 mm deep	C^b (%)
100—75	100—90	0.01—0.1
70—45	85—75	0.2—0.4
40—20	70—40	0.3—0.9

[a] Where effective litter cover is less than 40% or canopy cover is less that 20%, use table 9. Also use Table 9 where woodlands are being grazed, harvested, or burned.

[b] The ranges in listed C-value are caused by the ranges in the specified forest litter and canopy covers and by variations in effective canopy heights.

Table 8
C-VALUES FOR MECHANICALLY PREPARED WOODLAND SITES

		Soil condition[b] and weed cover[c] (%)							
	Mulch cover[a]	Excellent		Good		Fair		Poor	
Site preparation	(%)	NC	WC	NC	WC	NC	WC	NC	WC
Disked, raked, or bedded[d]	None	52	20	72	27	85	32	94	36
	10	33	15	46	20	54	24	60	26
	20	24	12	34	17	40	20	44	22
	40	17	11	23	14	27	17	30	19
	60	11	8	15	11	18	14	20	15
	80	5	4	7	6	9	8	10	9
Burned[e]	None	25	10	26	10	31	12	45	17
	10	23	10	24	10	26	11	36	16
	20	19	10	19	10	21	11	27	14
	40	14	9	14	9	15	9	17	11
	60	8	6	9	7	10	8	11	8
	80	4	4	5	4	5	4	6	5
Drum chopped[f]	None	16	7	17	7	20	8	29	11
	10	15	7	16	7	17	8	23	10
	20	12	6	12	6	14	7	18	9
	40	9	6	9	6	10	6	11	7
	60	6	5	6	5	7	5	7	5
	80	3	3	3	3	3	3	4	4

[a] Percentage of surface covered by residue in contact with the soil.

[b] Excellent soil condition — highly stable soil aggregates in topsoil with fine tree roots and litter mixed in; good — moderately stable soil aggregates in topsoil or highly stable aggregates in subsoil (topsoil removed during raking), only traces of litter mixed in; fair — highly unstable soil aggregates in topsoil or moderately stable aggregates in subsoil, no litter mixed in; poor — no topsoil, highly erodible soil aggregates in subsoil, no litter mixed in.

[c] NC, no live vegetation; WC, 75% cover of grass and weeds having an average drop fall height of 500 mm. For intermediate percentages of cover, interpolate between columns.

[d] Modify the listed C-values as follows to account for effects of surface roughness and aging: first year after treatment, multiply listed C-values by 0.50 for rough surface (depressions > 150 mm), by 0.65 for moderately rough, and by 0.90 for smooth (depressions < 50 mm); for 1 to 4 years after treatment, multiply listed factors by 0.7; for 4 + to 8 years, use Table 9; more than 8 years, use Table 7.

[e] For first 3 years, use C-values as listed; for 3 + to 8 years after treatment, use Table 9; more than 8 years after treatment, use Table 7.

Table 9
C-VALUES FOR PERMANENT PASTURE, RANGE, AND IDLE LAND[a]

Vegetative canopy			Cover that contacts the soil surface						
				Percent ground cover					
Type and height[b]	Percent cover[c]	Type[d]	0	20	40	60	80	95 +	
No appreciable		G	45	20	10	4.2	1.3	0.3	
canopy		W	45	24	15	9.1	4.3	1.1	
Tall weeds or short	25	G	36	17	9	3.8	1.3	0.3	
brush with average		W	36	20	13	8.3	4.1	1.1	
drop fall height of	50	G	26	13	7	3.5	1.2	0.3	
500 mm		W	26	16	11	7.6	3.9	1.1	
	75	G	17	10	6	3.2	1.1	0.3	
		W	17	12	9	6.8	3.8	1.1	
Appreciable brush or	25	G	40	18	9	4.0	1.3	0.3	
bushes, with aver-		W	40	22	14	8.7	4.2	1.1	
age drop fall height	50	G	34	16	8	3.8	1.2	0.3	
of 2 m		W	34	19	13	8.2	4.1	1.1	
	75	G	28	14	8	3.6	1.2	0.3	
		W	28	17	12	7.8	4.0	1.1	
Trees, but no appre-	25	G	42	19	10	4.1	1.3	0.3	
ciable low brush;		W	42	23	14	8.9	4.2	1.1	
average drop fall	50	G	39	18	9	4.0	1.3	0.3	
height		W	39	21	14	8.7	4.2	1.1	
	75	G	36	17	9	3.9	1.3	0.3	
		W	36	20	13	8.4	4.1	1.1	

[a] Vegetation and mulch are assumed to be randomly distributed.
[b] Height of canopy is the average distance drops fall from the canopy to the ground. Canopy effect is negligible where fall height exceeds 10 m.
[c] Percentage of the total land area covered by canopy.
[d] G, grass or litter, at least 50 mm deep; W, broadleaf herbaceous plants, undecayed residues, or both.

Table 10
P-VALUES AND SLOPE-LENGTH LIMITS FOR CONTOURING

Land slope (%)	P-value	Maximum length[a] (m)
1—2	0.60	120
3—5	0.50	90
6—8	0.50	60
9—12	0.60	40
13—16	0.70	25
17—20	0.80	20
21—25	0.90	15

[a] Limit may be increased by 25% if residue cover after crop seedlings will regularly exceed 50%.

Adapted from Wischmeier and Smith.[3]

Table 11

P-VALUES, MAXIMUM STRIP WIDTHS, AND SLOPE-LENGTH LIMITS FOR CONTOUR STRIP CROPPING

Land slope (%)	P-values[a]			Strip width[b] (m)	Maximum length (m)
	A	B	C		
1—2	0.30	0.45	0.60	40	240
3—5	0.25	0.38	0.50	30	180
6—8	0.25	0.38	0.50	30	120
9—12	0.30	0.45	0.60	25	70
13—16	0.35	0.52	0.70	25	50
17—20	0.40	0.60	0.80	20	35
21—25	0.45	0.68	0.90	15	30

[a]   P-values: (A) For 4-year rotation of row crop, small grain with meadow seeding, and 2 years of meadow. A second row crop can replace the small grain if meadow is established in it. (B) For 4-year rotation of 2 years row crop, winter grain with meadow seeding, and 1-year meadow. (C) For alternate strips of row crop and small grain.

[b]   Adjust strip-width limit, generally downward, to accommodate widths of farm equipment.

Adapted from Wischmeier and Smith.[3]

Table 12

P-VALUES FOR CONTOUR-FARMED TERRACED FIELDS[a]

Land slope (%)	Farm planning	
	Contour factor[b]	Stripcrop factor
1—2	0.60	0.30
3—8	0.50	0.25
9—12	0.60	0.30
13—16	0.70	0.35
17—20	0.80	0.40
21—25	0.90	0.45

[a]   Slope length is the horizontal terrace interval. The listed values are for contour farming. No additional contouring factor is used in the computation.

[b]   Use these values for control of inter-terrace erosion within specified soil loss tolerances.

Adapted from Wischmeier and Smith.[3]

Table 13

PERCENTAGE OF THE AVERAGE ANNUAL RAINFALL FACTOR WHICH NORMALLY OCCURS BETWEEN JANUARY 1 AND THE INDICATED DATES[a] COMPUTED FOR THE GEOGRAPHIC AREAS SHOWN IN FIGURE 2

Area no.	Jan. 1	Jan. 15	Feb. 1	Feb. 15	Mar. 1	Mar. 15	Apr. 1	Apr. 15	May 1	May 15	June 1	June 15	July 1	July 15	Aug. 1	Aug. 15	Sept. 1	Sept. 15	Oct. 1	Oct. 15	Nov. 1	Nov. 15	Dec. 1	Dec. 15
1	0	0	0	0	0	0	1	2	3	6	11	23	36	49	63	77	90	95	98	99	100	100	100	100
2	0	0	0	0	1	1	2	3	6	10	17	29	43	55	67	77	85	91	96	98	99	100	100	100
3	0	0	0	0	1	1	2	3	6	13	23	37	51	61	69	78	85	91	94	96	98	99	99	100
4	0	0	1	1	2	3	4	7	12	18	27	38	48	55	62	69	76	83	90	94	97	98	99	100
5	0	1	2	3	4	6	8	13	21	29	37	46	54	60	65	69	74	81	87	92	95	97	98	99
6	0	0	0	0	1	1	1	2	6	16	29	39	46	53	60	67	74	81	88	95	99	99	100	100
7	0	1	1	2	3	4	6	8	13	25	40	49	56	62	67	72	76	80	85	91	97	98	99	99
8	0	1	3	5	7	10	14	20	28	37	48	56	61	64	68	72	77	81	86	89	92	95	98	99
9	0	2	4	6	9	12	17	23	30	37	43	49	54	58	62	66	70	74	78	82	86	90	94	97
10	0	1	2	4	6	8	10	15	21	29	38	47	53	57	61	65	70	76	83	88	91	94	96	98
11	0	1	3	5	7	9	11	14	18	27	35	41	46	51	57	62	68	73	79	84	89	93	96	98
12	0	0	0	0	1	1	2	3	5	9	15	27	38	50	62	74	84	91	95	97	98	99	99	100
13	0	0	0	1	1	2	3	5	7	12	19	33	48	57	65	74	82	88	93	96	98	99	100	100
14	0	0	0	1	2	3	4	6	9	14	20	28	39	52	63	72	80	87	91	94	97	98	99	100
15	0	0	1	2	3	4	6	8	11	15	22	31	40	49	59	69	78	85	91	94	96	98	99	100
16	0	1	2	3	4	6	8	10	14	18	25	34	45	56	64	72	79	84	89	92	95	97	98	99
17	0	1	2	3	4	5	6	8	11	15	20	28	41	54	65	74	82	87	92	94	96	97	98	99
18	0	1	2	4	6	8	10	13	19	26	34	42	50	58	63	68	74	79	84	89	93	95	97	99
19	0	1	3	6	9	12	16	21	26	31	37	43	50	57	64	71	77	81	85	88	91	93	95	97
20	0	2	3	5	7	10	13	16	19	23	27	34	44	54	63	72	80	85	89	91	93	95	96	98
21	0	3	6	10	13	16	19	23	26	29	33	39	47	58	68	75	80	83	86	88	90	92	95	97
22	0	3	6	9	13	17	21	27	33	38	44	49	55	61	67	71	75	78	81	84	86	90	94	97
23	0	3	5	7	10	14	18	23	27	31	35	39	45	53	60	67	74	80	84	86	88	90	93	95
24	0	3	6	9	12	16	20	24	28	33	38	43	50	59	69	75	80	84	87	90	92	94	96	98
25	0	1	3	5	7	10	13	17	21	24	27	33	40	46	53	61	69	78	89	92	94	95	97	98
26	0	2	4	6	8	12	16	20	25	30	35	41	47	56	67	75	81	85	87	89	91	93	95	97
27	0	1	2	3	5	7	10	14	18	22	27	32	37	46	58	69	80	89	93	94	95	96	97	99
28	0	1	3	5	7	9	12	15	18	21	25	29	36	45	56	68	77	83	88	91	93	95	97	99
29	0	1	2	3	4	5	7	9	11	14	17	22	31	42	54	65	74	83	89	92	95	97	98	99
30	0	1	2	3	4	5	6	8	10	14	19	26	34	45	56	66	76	82	86	90	93	95	97	99
31	0	0	0	1	2	3	4	5	7	12	17	24	33	42	55	67	76	83	89	92	94	96	98	99
32	0	1	2	3	4	5	6	8	10	13	17	22	31	42	52	60	68	75	80	85	89	92	96	98
33	0	1	2	4	6	8	11	13	15	18	21	26	32	38	46	55	64	71	77	81	85	89	93	97

[a]   For dates not listed in the table, interpolate between adjacent values.

Adapted from Wischmeier and Smith.[3]

DESIGN PROBABILITIES

Probabilities of annual rainfall factors exceeding certain values are available for 5, 20, and 50% probabilities for 181 U.S. locations.[3] Also available are the magnitude of single-storm rainfall factors normally exceeded every year and once in 2, 5, 10, and 20 years. Enough data are available so that reasonable estimates can be made for most areas of the U.S. These data can be used to replace the average rainfall factor values so that erosion amounts can also be associated with probabilities, either on an annual or storm basis.

FIGURE 2.   Key map selection of rainfall factor distribution from Table 13. (Redrawn from Wischmeier and Smith.[3])

SEDIMENT YIELDS

The USLE is not an estimator of downstream sediment yields. It is applicable only to areas that are somewhat similar to the areas from where the data were derived for determination of relationships used. These areas, for the most part, were plots of narrow width (usually less than 5 m) and limited length (usually less than 90 m). Slopes were usually uniform, and there were no gullies. Generally, deposition of sediments detached on the plot did not occur within the plot. Hence, on a watershed where channel erosion or deposition occurs, the USLE would not be suitable for estimating sediment yields. A model for estimating sediment and chemical yields from field-sized areas that have channel erosion and deposition has been developed.[13] It also includes a hydrologic component, so it is suitable for individual storm soil loss estimation. However, it is much more complex than the USLE, requiring the use of a digital computer and greatly detailed data inputs, at least as compared to the USLE.

ACCURACY OF PREDICTION

The USLE has received very little independent evaluation, and its accuracy in predicting soil erosion is largely a matter of speculation. However, it has been found to be a very useful tool in evaluating alternatives for land use, management, and treatment.

While the USLE can be used to predict soil erosion for any period of time, ranging from a single storm to a long-term average, it was developed primarily for use in estimating long-term average annual soil loss from cropped agricultural lands. It is recognized that for short-time periods, predictions can be very poor because of the wide year-to-year and day-to-day variation in factors affecting soil erosion that cannot be accounted for by using average crop period or annual C-values. For further discussions, the reader is referred to Wischmeier's writings on the use and misuse of the USLE.[14]

Table 14
GENERALIZED VALUES OF THE COVER AND
MANAGEMENT FACTOR: ANNUAL C-VALUES[a]

Line no.	Crop, rotation, and management[c]	C-value productivity level[b] (%)	
		High	**Mod.**
—	Continuous fallow, tilled up and down slope	100	100
Corn			
1	C, RdR, fall TP, conv (1)	54	62
2	C, RdR, spring TP, conv (1)	50	59
3	C, RdL, fall TP, conv (1)	42	52
4	C, RdR, wc seeding, spring TP, conv (1)	40	49
5	C, RdL, standing, spring TP, conv (1)	38	48
6	C, fall shred stalks, spring TP, conv (1)	35	44
7	C(silage)-W(RdL, fall TP) (2)	31	35
8	C, RdL, fall chisel, spring disk, 40—30% rc (1)	24	30
9	C(silage), W wc seeding, no-till pl in c-k W (1)	20	24
10	C(RdL)-W(RdL, spring TP) (2)	20	28
11	C, fall shred stalks, chisel pl, 40—30% rc (1)	19	26
12	C-C-C-W-M, RdL, TP for C, disk for W (5)	17	23
13	C, RdL, strip till row zones, 55—40% rc (1)	16	24
14	C-C-C-W-M-M, RdL, TP for C, disk for W (6)	14	20
15	C-C-W-M, RdL, TP for C, disk for W (4)	12	17
16	C, fall shred, no-till Pl, 70—50% rc (1)	11	18
17	C-C-W-M-M, RdL, TP for C, disk for W (5)	8.7	14
18	C-C-C-W-M, RdL, no-till pl 2d and 3rd C (5)	7.6	13
19	C-C-W-M, RdL, no-till pl 2d C (4)	6.8	11
20	C, no-till pl in c-k wheat, 90—70% rc (1)	6.2	14
21	C-C-C-W-M-M, no-till pl 2nd and 3rd C (6)	6.1	11
22	C-W-M, RdL, TP for C, disk for W (3)	5.5	9.5
23	C-C-W-M-M, RdL, no-till pl 2d C (5)	5.1	9.4
24	C-W-M-M, RdL, TP for C, disk for W (4)	3.9	7.4
25	C-W-M-M-M, RdL, TP for C, disk for W (5)	3.2	6.1
26	C, no-till pl in c-k sod, 95—80% rc (1)	1.7	5.3
Cotton[d]			
27	Cot, conv (western plains) (1)	42	49
28	Cot, conv (south) (1)	34	40
Meadow			
29	Grass and legume mix	0.4	1
30	Alfalfa, lespedeza or sericia	2.0	—
31	Sweet clover	2.5	—
Sorghum, grain (western plains)[d]			
32	RdL, spring TP, conv (1)	43	53
33	No-till pl in shredded 70—50% rc	11	18
Soybeans[d]			
34	B, RdL, spring TP, conv (1)	48	54
35	C-B, TP annually, conv (2)	43	51
36	B, no-till pl	22	28
37	C-B, no-till pl, fall shred C stalks (2)	18	22
Wheat			
38	W-F, fall TP after W (2)	38	—
39	W-F, stubble mulch, 600 kg/ha rc (2)	32	—
40	W-F, stubble mulch, 1100 kg/ha rc (2)	21	—
41	Spring W, RdL, Sept. TP, conv (North and South Dakota) (1)	23	—
42	Winter W, RdL, Aug. TP, conv (Kansas) (1)	19	—
43	Spring W, stubble mulch, 800 kg/ha rc (1)	15	—

Table 14 (continued)
GENERALIZED VALUES OF THE COVER AND
MANAGEMENT FACTOR: ANNUAL C-VALUES[a]

Line no.	Crop, rotation, and management[c]	C-value productivity level[b] (%)	
		High	Mod.
44	Spring W, stubble mulch, 1400 kg/ha rc (1)	12	—
45	Winter W, stubble mulch, 800 kg/ha rc (1)	11	—
46	Winter W, stubble mulch, 1400 kg/ha rc (1)	10	—
47	W-M, conv (2)	5.4	—
48	W-M-M, conv (3)	2.6	—
49	W-M-M-M, conv (4)	2.1	—

Note: B, soybeans; C, corn; c-k, chemically killed; conv, conventional; cot, cotton; F, fallow; M, grass-and-legume hay; pl, plant; W, wheat; wc, winter cover; rc, weight of crop residue per hectare remaining on surface after new crop seeding; % rc, percentage of soil surface covered by residue mulch after new crop seeding; 70 to 50% rc, 70% cover for C-values in first column and 50% for second column; RdR, residues (corn stover, straw, etc.) removed or burned; RdL, all residues left on field (on surface or incorporated); TP, turn plowed (upper 120 or more mm of soil inverted, covering residues).

[a] These generalized values show approximately annual C-values for various crop systems, but locationally derived C-values that differ with rainfall pattern and planting dates should be used for conservation planning.

[b] High level is exemplified by long-term yield averages greater than 2 t/ha corn or 7 t/ha grass-and-legume hay or cotton management that regularly provides good stands and growth.

[c] Numbers in parentheses indicate number of years in the rotation cycle. Number (1) designates a continuous one-crop system.

[d] Grain sorghum, soybeans, or cotton may be substituted for corn in lines 12, 14, 15, 17 to 19, and 21 to 25 to estimate C-values for sod-based rotations.

Adapted from Stewart et al.[12]

Table 15
AN EXAMPLE OF THE DETERMINATION OF A ROTATION C-VALUE

Date	Event	Period	Table no.	Line no.	C_i	f_i
Nov. 1	Chisel	SB	1	34	0.10	0.15
Apr. 15	Secondary tillage	SB	1	42	0.17	0.04
May 1	Plant soybeans	SB	1	42	0.17	0.11
June 1	10% canopy	1	1	42	0.15	0.20
July 1	50% canopy	2	1	42	0.13	0.11
July 15	75% canopy	3	1	42	0.10	0.33
Oct. 1	Harvest (96% canopy)	4	4	(80% cover)	0.09	0.25
May 1	No-till plant corn	SB	1	120	0.25	0.11
June 1	10% canopy	1	1	120	0.20	0.20
July 1	50% canopy	2	1	120	0.19	0.11
July 15	75% canopy	3	1	120	0.11	0.33
Oct. 1	Harvest (96% canopy)	4	4	(95% cover)	0.03	0.06
Nov. 1	Chisel					2.00 total

Note: $\Sigma f_i C_i = 0.266$; avg. annual C $= 0.133$

SUMMARY

The use of the USLE for water erosion caused by rainfall has been presented. The discussion, of necessity, has been brief. Considerable judgment is needed to apply the equation and to determine the factor values to use. The reader should consult the references, particularly Wischmeier and Smith.[3]

The USLE, because it reflects the impact of the major factors on soil erosion, will likely be in use for some time. It was specifically developed to estimate long-term average annual erosion so that alternatives for using, managing, and treating agricultural lands could be evaluated.

Coefficient and factor evaluation is an ongoing process as new farming methods and new land treatments are developed and additional data become available. An active international research program regularly provides improved coefficient values, factor relationships, and new applications.

REFERENCES

1. **Troeh, F. R., Hobbs, J. A., and Donahue, R. L.,** *Soil and Water Conservation for Productivity and Protection,* Prentice-Hall, Englewood Cliffs, N.J., 1980, chap. 2.
2. **Wischmeier, W. H. and Smith, D. D.,** Rainfall energy and its relationship to soil loss, *Trans. Am. Geophys. U.,* 39, 285, 1958.
3. **Wischmeier, W. H. and Smith, D. D.,** Predicting Rainfall Erosion Losses, Agriculture Handbook 537, U.S. Department of Agriculture, Washington, D.C., 1978.
4. **Foster, G. R., McCool, D. K., Renard, K. G., and Moldenhauer, W. C.,** Conversion of the Universal Soil Loss Equation to SI metric units, *J. Soil Water Conserv.,* 36, 355, 1981.
5. **Olson, T. C. and Wischmeier, W. H.,** Soil erodibility evaluations for soils on the runoff and erosion stations, *Proc. Soil Sci. Soc. Am.,* 27, 590, 1963.
6. **Wischmeier, W. H., Johnson, C. B., and Cross, B. V.,** A soil erodibility nomograph for farmland and construction sites, *J. Soil Water Conserv.,* 26, 189, 1971.
7. **Wischmeier, W. H.,** Cropping-management factor evaluation for a Universal Soil Loss Equation, *Proc. Soil Sci. Soc. Am.,* 23, 322, 1960.
8. **Meyer, L. D. and Ports, M. A.,** Prediction and Control of Urban Erosion and Sedimentation, in Proc. Natl. Symp. Urban Hydrology, Hydraulics and Sedimentation, University of Kentucky, Lexington, 1976, 323.
9. **Dissmeyer, G. E. and Foster, G. R.,** A Guide for Predicting Sheet and Rill Erosion on Forest Land, Tech. Publ. SA-TP11, U.S. Department of Agriculture, Washington, D.C., 1980.
10. **Dissmeyer, G. E. and Foster, G. R.,** Estimating the cover-management factor (C) in the Universal Soil Loss Equation for forest conditions, *J. Soil Water Conserv.,* 36, 235, 1981.
11. **Wischmeier, W. H.,** Estimating the Soil Loss Equation Cover and Management Factor for Undisturbed Areas, Proc. Sediment Yield Workshop, ARS-S-40, U.S. Department of Agriculture, Washington, D.C., 1975.
12. **Stewart, B. A., Woolhiser, D. A., Wischmeier, W. H., Caro, J. H., and Frere, M. H.,** Control of Water Pollution from Cropland, Vol. 1, ARS-H-5-1, U.S. Department of Agriculture, Washington, D.C., 1975.
13. **Foster, G. R., Lane, L. H., Nowlin, J. D., Laflen, J. M., and Young, R. A.,** Estimating erosion and sediment yield on field-sized areas, *Trans. Am. Soc. Agric. Eng.,* 24, 1253, 1981.
14. **Wischmeier, W. H.,** Use and misuse of the Universal Soil Loss Equation, *J. Soil Water Conserv.,* 31, 5, 1976.

DESIGNING GRASS WATERWAYS FOR CHANNEL STABILIZATION AND SEDIMENT FILTRATION

B. J. Barfield and J. C. Hayes

GENERAL CONSIDERATIONS

Uses of Vegetated Waterways

Vegetated waterways have been used for many years to prevent channel erosion resulting from storm runoff in intermittent drainage channels. When properly installed, the vegetation dissipates most of the energy resulting from shear forces on the channel bed, thus protecting the soil surface itself from erosion. In addition, the root structure of the vegetation enhances the ability of the soil to withstand the shearing forces of runoff while remaining intact.

Since persistent flooding will normally cause death for most vegetation, these waterways are limited to channels that do not experience base flow. Typical applications include stabilization of terrace outlets, gulley stabilization,[13] road ditch stabilization, and stabilization of surface drains. In applications where low levels of base flow occur, a small ditch lined with riprap or other stabilizers can sometimes be placed in the center of the ditch with sufficient capacity to handle the base flow.

Another application of grass waterways is for the filtration of sediment from overland flow. In this application, the vegetation is not typically submerged by the flow. However, the vegetation does retard the flow, reducing flow velocities and sediment transport capacity as well as increasing the infiltration opportunity time. When designed primarily as a sediment filter, the channels should have a wide, flat bottom in order to promote shallow flow throughout the filter.

Selecting An Acceptable Vegetation

Care must be exercised in selecting an acceptable vegetation for grass waterways in order to assure that the vegetation characteristics match the characteristics necessary for stabilizing the waterway. Vegetative species that are of uniform density are more desirable than those that tend to form clumps, since the bare soil between the clumps will not be stabilized.

In order to assure proper vegetal establishment and survival during climatic extremes, the proper vegetative types must be matched to the climatic regions. A list is given in Table 1 of vegetative species recommended for different climatic regions given in Figure 1.

Other factors to be considered in selection of the appropriate vegetative species are the speed with which vegetation can be established vs. the ease with which flow can be diverted, potential uses of the vegetation, duration and velocity of the flow, and the ability of the vegetation to stabilize the soil in a mowed condition.

DESIGNING WATERWAYS FOR CHANNEL STABILIZATION

Selecting a Design Storm

The functional requirement of a waterway is typically that it convey the design storm without eroding the channel or overtopping. Once a design storm is selected, the channel shape can be determined to meet these requirements.

Guidelines for determining the magnitude of the design storm will sometimes be dictated by a regulatory authority. In this case, the regulatory authority will typically specify a return period rainfall. If no guidelines are available, a risk analysis approach can be used to specify a return period storm. By assuming that rare events occur as a Bernoulli process, it can be shown[5] that the probability of one or more exceedences of a T year event in n years is

Table 1
VEGETATION RECOMMENDED FOR GRASSED WATERWAYS

Geographical area of U.S.[a]	Vegetation
Northeastern — L, N, R, S	Kentucky bluegrass, red top, tall fescue, white clover
Southeastern — N, O, P, T, U	Kentucky bluegrass, tall fescue, Bermuda, brome, Reed canary
Upper Mississippi — K, L, M, N	Brome, Reed canary, tall fescue, Kentucky blue grass
Western Gulf — H, I, J, P, T	Bermuda, King Ranch bluestem, native grass mixture, tall fescue
Southwestern — C, D, G	Intermediate wheatgrass, western wheatgrass, smooth brome, tall wheatgrass
Northern Great Plains — G, F, H	Smooth brome, western wheatgrass, red top switchgrass, native bluestem mixture

[a]   Letters refer to land resource regions shown in Chapter 1, but recommended vegetation does not necessarily apply to all areas in the region.

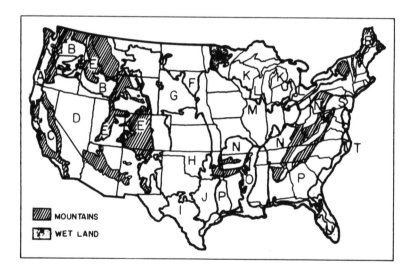

FIGURE 1.   Major land resource regions of the U.S. (A) Northwestern forest, forage, and specialty crop region. (B) Northwestern wheat and range region. (C) California subtropical fruit, truck, and specialty crop region. (D) Western range and irrigated region. (E) Rocky Mountain range and forest region. (G) Western Great Plains range and irrigated region. (H) Central Great Plains winter wheat and range region. (I) Southwestern plateaus and plains, range, and cotton region. (J) Southwestern prairies, cotton and forage region. (K) Northern lake states, forest and forage region. (L) Lake states fruit, truck, and dairy region. (M) Central feed grains and livestock region. (N) East and Central general farming and forest region. (O) Mississippi Delta cotton and feed grains region. (P) South Atlantic and Gulf Slope cash crop, forest, and livestock region. (R) Northeastern forage and forest region. (S) Northern Atlantic Slope truck, fruit, and poultry region. (T) Atlantic and Gulf Coast lowlands, forest and truck crop region. (U) Florida subtropical fruit, truck crop, and range region.

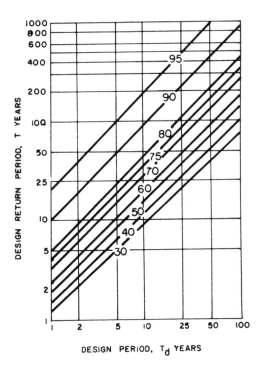

FIGURE 2.   Design return period required as a function of design life to be given a percent confident (curve parameter) that the design condition is not exceeded.

$$f(t,n) = 1 - (1 - 1/T)^n \qquad (1)$$

The relationship is shown graphically in Figure 2. For example, if one is constructing a vegetated channel with a design life of 10 years and wants to be 90% sure that the design storm is not exceeded, then $f(T,n)$ is 0.10 and n is 10. Equation 1 can be solved by trial and error to find a return period T of 95 years. This can also be determined graphically from Figure 2.

Once the design rainfall has been determined, a design runoff can be calculated from one of several techniques presented in Section 5 of this handbook. These runoff values are used to determine a design cross section. Typically, the peak discharge in a storm is used as the design discharge. If the waterway is exceptionally long, it may be desirable to calculate different peaks discharges for several different sections. If the waterway is short, the peak discharge for the entire watershed draining through the channel is the appropriate design flow rate.

Waterway Shapes

The typical cross-sectional shape of a grassed waterway will either be trapezoidal, parabolic, or triangular. Geometric properties of these channel shapes are given in Figure 3.

For unlined channels, the triangular and trapezoidal channel shape tends to become a parabolic section due to erosion and deposition; therefore, the parabolic section is typically recommended for stability. A vegetated waterway is designed to be stable at all points along the boundary; therefore, other factors should dictate channel shape. In the case of heavy sediment loads in the inflow to the channel shape, a parabolic section would still be recommended in order to make deposition occur uniformly.

The following factors should be considered in the selection of a channel cross section.

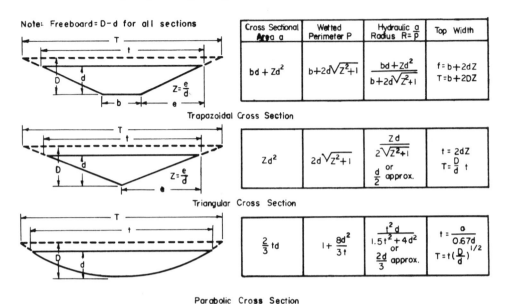

Cross Sectional Area a	Wetted Perimeter P	Hydraulic Radius R=a/P	Top Width
$bd + Zd^2$	$b+2d\sqrt{Z^2+1}$	$\dfrac{bd + Zd^2}{b+2d\sqrt{Z^2+1}}$	$t = b+2dZ$ $T = b+2DZ$
Zd^2	$2d\sqrt{Z^2+1}$	$\dfrac{Zd}{2\sqrt{Z^2+1}}$ or $\dfrac{d}{2}$ approx.	$t = 2dZ$ $T = \dfrac{D}{d}t$
$\dfrac{2}{3}td$	$1+\dfrac{8d^2}{3t}$	$\dfrac{t^2d}{1.5t^2+4d^2}$ or $\dfrac{2d}{3}$ approx.	$t = \dfrac{a}{0.67d}$ $T = t\left(\dfrac{D}{d}\right)^{1/2}$

Note: Freeboard = D−d for all sections

$Z = \dfrac{e}{d}$

Trapezoidal Cross Section

$Z = \dfrac{e}{d}$

Triangular Cross Section

Parabolic Cross Section

FIGURE 3. Properties of typical channels. (Redrawn from Schwab, G. O., Frevert, R. K., Edminster, T. W., and Barnes, K. K., *Soil and Water Conservation Engineering*, John Wiley & Sons, New York, 1966. With permission.)

1. A triangular cross section is best suited for blade-type equipment unless a trapezoidal section is selected with a bottom width greater than the blade width.
2. Side slopes of 4:1 (4 horizontal to 1 vertical) or greater are necessary for ease in maintenance and equipment crossing.
3. Excavation is minimized for shallow broad-bottomed trapezoidal channels as compared to triangular channels; however, deposition can occur in low flows in the wide channels, resulting in a potential for meandering. At high flows, the meanders can cause localized high levels of turbulence and damage to the vegetation.

Calculating Channel Velocities

Several procedures have been presented for calculating flow velocities in vegetated waterways. These procedures range in complexity from simple regression relationships to complex theoretical relationships.[3,18,24] The procedure that is most commonly used is the Manning equation:

$$V = \frac{1}{n} R^{2/3}S^{1/2} \qquad (2)$$

where V is the mean velocity, n is Manning's roughness, S is channel slope, and R is the hydraulic radius given by

$$R = \frac{A}{P} \qquad (3)$$

where A is the cross-sectional area and P is the wetted perimeter. Equations for A, R, and P are given in Figure 3.

The difficulty with the use of Equation 2 for grassed waterways is that Manning's n is not constant, but varies widely with vegetative types, flow depths, and height of vegetation.

Typically, Manning's n increases with flow depth until the vegetation is submerged, and then decreased with depth.

Based on tests by Ree,[14-17] techniques have been developed to predict roughness values as a function of vegetative types and heights of vegetation. Ree divided vegetation into five retardance classes, as shown in Table 2, based on its resistance to flow. For each retardance class, curves of Manning's n vs. the product of velocity times hydraulic radius were developed. The average curves for each retardance class are presented in Figure 4 as a function of n vs. flow Reynold's number rather than the VR product in order to be dimensionless. (For simplicity in the conversion, v was assumed to be 10^{-5} ft^2/sec or 9.3×10^{-7} m^2/5.)

Temple[21] proposed that the relationship between Reynold's number and n could be defined within 10% accuracy by

$$n = \exp(0.01329 \, C_I \, [\ln(R_v)]^2 - 0.09543 \, C_I \, \ln(R_v) + 0.2971 \, C_I - 4.16) \qquad (4)$$

where C_I is the retardance curve coefficient given in Table 2 and R_v is the normalized Reynold's number given by

$$R_v = R \times 10^{-5} \qquad (5)$$

Limits on the applicability of Equation 4 are

$$n \leqslant (C_I + 1)/36$$

$$0.1 < R_v < 36$$

$$0 < C_I < 10$$

A plot of Equation 4 is also given in Figure 4 for each retardance class, indicating an excellent fit to the experimental curves within the ranges proposed.

By using Equation 4 or Figure 4 for n along with the Manning equation, the flow velocity can be calculated for a given retardance class and depth of flow. In order to eliminate the need for successive iterations required by the variable n vs. Reynold's number relationship, nomographs have been prepared, as shown in Figure 5, relating velocity to hydraulic radius and slope.

Designing the Channel for Stability and Capacity

The first channel design consideration should be for stability of the cross section, assuming that the design storm occurs when the channel is in its most vulnerable state. This would typically be under mowed conditions in which a maximum amount of bare soil would be exposed. After designing for stability, the channel capacity should be increased to allow for the possibility of the design storm occurring when the vegetation is in the unmowed condition in which the retardance class would typically be higher. A free board of 0.1 to 0.15 times the depth of flow is typically added for safety. A schematic of the design flow depths and geometrics under different retardance is given in Figure 6.

Permissible velocities are given in Table 3 for designing a vegetated channel to be stable, i.e., no erosion. It is typically assumed that these design criteria will be applied when the vegetation has been mowed, i.e., at its minimum retardance. Design procedures are illustrated in the following example.

Example Problem No. 1

Design a Bermudagrass-lined channel to convey 0.708 m^3/sec (25 cfs) on a 4% slope.

Table 2
VEGETAL RETARDANCE CLASSES

Retardance	Retardance curve coef. C_1	Cover	Condition	Avg height of vegetation	
				cm	in.
A	10.00	Reed canarygrass	Excellent stand, tall	91	36
		Yellow bluestem Ischaemum	Excellent stand, tall	91	36
B	7.64	Smooth bromegrass	Good stand, mowed	30—38	12—15
		Bermudagrass	Good stand, tall	30	12
		Native grass mixture (little bluestem, blue grama, and other long and short Midwest grasses)	Good stand, unmowed	—	—
		Tall fescue	Good stand, unmowed	41	18
		Lespedeza sericea	Good stand, not woody	48	19
		Grass-legume mixture — timothy, smooth bromegrass, or orchardgrass	Good stand, uncut	51	20
		Reed canarygrass	Good stand, mowed	33	12—15
		Tall fescue, with birdsfoot trefoil or lodino	Good stand, uncut	41	18
		Blue grama	Good stand, uncut	33	13
C	5.60	Bahia	Good stand, uncut	18	7
		Bermudagrass	Good stand, mowed	15	6
		Redtop	Good stand, headed	38—51	15—20
		Grass-legume mixture — summer (orchard-grass, redtop, Italian ryegrass, and common lespedeza)	Good stand, uncut	15—20	6—8
		Centipedegrass	Very dense cover	15	6
		Kentucky bluegrass	Good stand, headed	15—30	6—12
D	4.44	Bermudagrass	Good stand, cut to 2.5-in. height	6	2.5
		Red fescue	Good stand, headed	30—46	12—18
		Buffalograss	Good stand, uncut	8—15	3—6

	Grass-legume mixture — fall, spring (orchardgrass, redtop, Italian ryegrass, and common lespedeza)	Good stand, uncut	10—12	4—5
	Lespedeza sericea	After cutting to 2-in. height, very good stand before cutting	5	2
E	2.88	Bermudagrass	Good stand, cut to 1.5-in. height	4	1.5
	Bermudagrass	Burned stubble	—	—

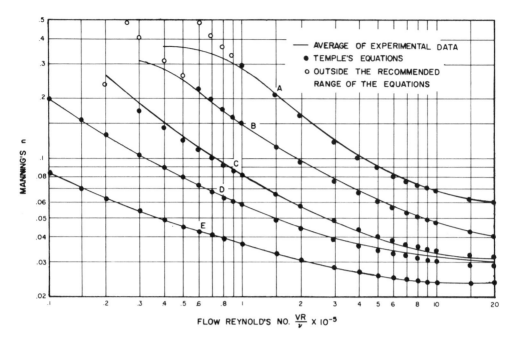

FIGURE 4. n-VR for various retardance classes.

The soil is easily eroded and the grass may be mowed to 6.35 cm (2.5 in.) or it may be uncut. Use a parabolic channel.

Solution:

Retardance B unmowed

Retardance D mowed

Permissible velocity 1.83 m/S (6 ft/8)

Design in mowed condition

$$A = Q/V = 0.708/1.83 = 0.386 \text{ m}^2(4.15 \text{ ft}^2)$$

From Figure 5 retardance class D

$$R = 0.21 \text{ m}(0.7 \text{ ft})$$

$$R = 0.21 = \frac{t^2 d}{1.5^2 + 4d^2}$$

$$A = 0.386 = \frac{2}{3} td$$

For small parabolic channels d ≃ 1.5R. Using this approximation, d = 0.32 m (1.03 ft)

$$d = 0.32 \text{ m}(1.03 \text{ ft})$$

$$t = 3A/2d = 3(0.386)/2(0.32) = 1.81 \text{ m}(5.94 \text{ ft})$$

check

$$R = \frac{(1.81)^2(0.32)}{1.5(1.81)^2 + 4(0.32)^2} = 0.197\text{m}(0.646 \text{ ft})$$

Retardance Class A

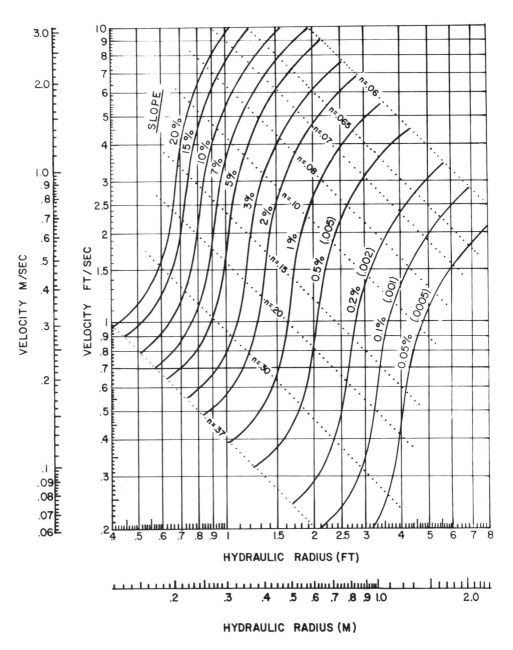

FIGURE 5A.   Solution for Manning's equation, vegetated waterways. Retardance class A.[20]

increase

$$d \text{ to } 0.365 \text{ m}(1.2 \text{ ft})$$

$$t = 3A/2d = 1.58 \text{ m}(5.21 \text{ ft})$$

$$R = 0.217 \text{ m}(0.715 \text{ ft})$$

Retardance Class B

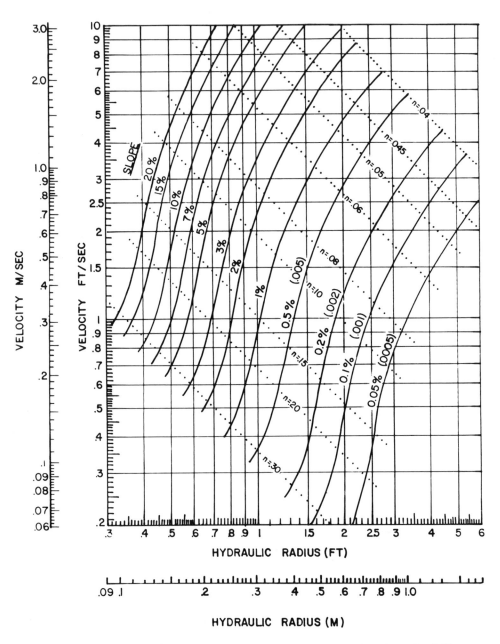

FIGURE 5B.   Solution for Manning's equation, vegetated waterways. Retardance class B.[20]

$$d = 0.357 \text{ m}(1.17 \text{ ft})$$

$$t = 3A/2d = 1.63 \text{ m}(5.35 \text{ ft})$$

$$R = 0.21 \text{ m}(0.7 \text{ ft}) \quad OK$$

The design for short condition is

Retardance Class C

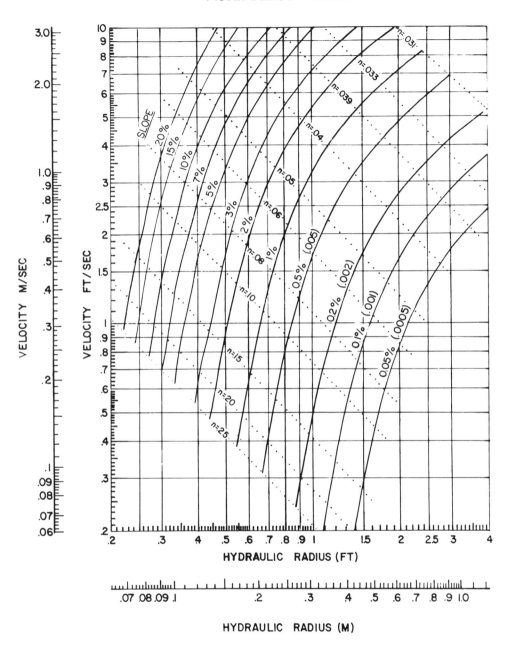

FIGURE 5C.   Solution for Manning's equation, vegetated waterways. Retardance class C.[20]

$$t = 1.63 \text{ m}(5.35 \text{ ft})$$

$$d = 0.357 \text{ m}(1.17 \text{ ft})$$

$$R = 0.21 \text{ m}(0.7 \text{ ft})$$

Now we must add depth using the same basic shape to get adequate capacity when the grass is long. When grass is long, the retardance class is B. Try

Retardance Class D

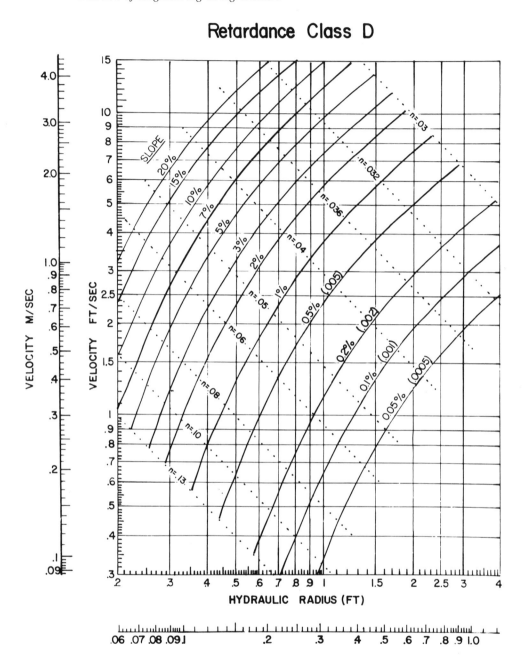

HYDRAULIC RADIUS (FT)

HYDRAULIC RADIUS (M)

FIGURE 5D.   Solution for Manning's equation, vegetated waterways. Retardance class D.[20]

$$d = 0.426 \text{ m}(1.40 \text{ ft})$$

New top width

$$t = 1.63\left(\frac{0.426}{0.357}\right)^{1/2} = 1.78 \text{ m}(5.85 \text{ ft})$$

$$R = \frac{t^2 d}{1.5t^2 + 4d^2} = 0.247 \text{ m}(0.81 \text{ ft})$$

Retardance Class E

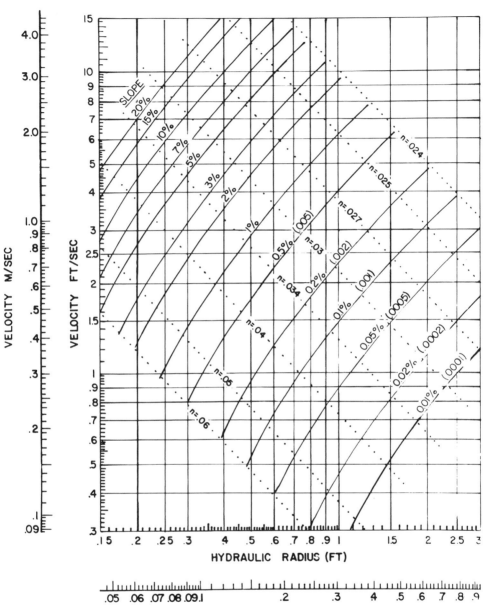

FIGURE 5E. Solution for Manning's equation, vegetated waterways. Retardance class E.[20]

From Figure 11 and retardance B with R = 0.247 m and S = 0.04 find V = 0.88 m/S (2.9 fps).

$$A = \frac{2}{3} t\, d = 0.507\ m^2 (5.46\ ft^2)$$

$$Q = VA = 0.88 \times 0.507 = 0.44\ m/S (15.8\ cfs)$$

too small

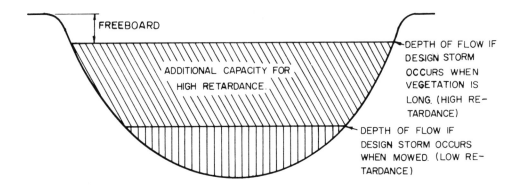

FIGURE 6. Schematic of the flow depths at low and high retardance. Note that when additional capacity is added, the channel geometry based on the low retardance design is maintained.

Table 3
PERMISSIBLE VELOCITIES FOR VEGETATED CHANNELS[14]

	Permissible velocity, m/S (fps)					
	Erosion-resistant soils (% slope)			Easily eroded soils (% slope)		
Cover	0—5	5—10	Over 10	0—5	5—10	Over 10
Bermudagrass	2.44 (8)	2.13 (7)	1.83 (6)	1.83 (6)	1.52 (5)	1.22 (4)
Buffalograss Kentucky bluegrass Smooth brome Blue grama Tall fescue	2.13 (7)	1.83 (6)	1.52 (5)	1.52 (5)	1.22 (4)	0.91 (3)
Lespedeza sericea Weeping lovegrass Kudzu Alfalfa Crabgrass	1.07 (3.5)	NR[a]	NR	0.76 (2.5)	NR	NR
Grass mixture	1.52 (5)	1.22 (4)	NR	1.22 (4)	.41 (3)	NR
Annuals for temporary protection	1.07 (3.5)	NR	NR	0.76 (2.5)	NR	NR

[a] Not recommended.

After Ree, W. O., *Agric. Eng.*, 30, 184, 1949. With permission.

try

$$d = 0.53 \text{ m}(1.75 \text{ ft})$$

$$t = 1.63\left(\frac{0.53}{0.357}\right)^{1/2} = 1.99 \text{ m}(6.54 \text{ ft})$$

$$R = 0.277 \text{ m}(0.91 \text{ ft})$$

$$V = 1.19 \text{ m/S}(3.9 \text{ fps})$$

$$A = 0.62 \text{ m}^2(6.68 \text{ ft}^2)$$

$$Q = 0.74 \text{ m}^3/\text{S}(26 \text{cfs}) \quad \text{OK}$$

add freeboard of 0.05 m(0.16 ft)

Table 4
APPROXIMATE CHANNEL DEPTH INCREMENT FOR
GRASSED WATERWAYS WITH RETARDANCE
INCREASE FROM CLASS C (SHORT) TO CLASS B (LONG)[10]

	Channel cross section and width					
Slope range (%)	Triangular (t = 2—6 m), trapezoidal (b = 2—6 m)		Parabolic (t = 6—27 m), trapezoidal (b = 6—27 m)		Trapezoidal (b = 27 m or more)	
1—2	0.17m	(0.6 ft)	0.15 m	(0.5 ft)	0.12 m	(0.4 ft)
2—5	0.13	(0.4)	0.12	(0.4)	0.09	(0.3)
5 or more	0.10	(0.3)	0.09	(0.3)	0.06	(0.2)

*   Adapted from Larson, C. L. and Manbeck, D. M., *Agric. Eng.*, 41, 694, 1960. With permission.

Final design (rounding to measurable distances)

$$d = .54 \text{ m}(1.8 \text{ ft})$$

$$t = 1.63 \left(\frac{0.54}{0.357} \right)^{1/2} = 2.0 \text{ m}(6.6 \text{ ft})$$

parabolic shape

As can be seen from the previous example, considerable trial and error is involved in determining the extra depth required to convey the design storm at a high retardance class. In order to eliminate some of the trial and error, Larson and Manbeck[10] compiled a first estimate of the incremental depth change required when going from a low to high retardance class. These values are tabulated in Table 4.

Construction and Maintenance Considerations
Construction of Waterways and Establishing Vegetation
   Construction procedures will depend heavily on the topography and equipment available. Small waterways may be constructed with farm equipment, but large waterways will require the use of dozers or earth-moving equipment. Waterways located in natural draws may require only minimal amounts of earth moving.
   Small gullies may be stabilized by simply reshaping the channel sides and vegetal establishment, but large gullies will typically require some type of grade stabilization structure.
   Before seeding the waterway, soil tests should be performed and the fertilizer and lime recommendations established. After bringing the soil to a proper fertility level, a mixture of quick growing annuals along with the recommended vegetation should be broadcast or drilled perpendicular to the direction of flow. Mulches applied to the surface will assist in stabilizing the soil and maintaining moisture necessary for vegetal establishment. A list of mulches, along with recommended rates, is given by Haan and Barfield.[6]

Maintenance
   Proper performance of a waterway requires both protection and maintenance of the vegetation. If a waterway is used as a roadway or animal pathway, failure often results due to channelized flow. Terraces that enter a waterway at a steep grade can also cause failure.

The optimum control of vegetation requires mowing and raking several times a year as well as annual fertilization. Sod breaks should be repaired.

Gwinn and Ree[4] evaluated the effects of lack of maintenance over a 32-year period on a triangular-shaped waterway of Bermudagrass. Over this period of years, the encroachment of trees and brush caused a reduction in channel capacity of 29% and a reduction in the permissible velocity of 27%. The retardance increased one class as a result of lack of maintenance; therefore, if the waterway will not be adequately maintained, the design for capacity should be increased at least one retardance class above the unmowed condition.

DESIGNING WATERWAYS FOR SEDIMENT FILTRATION

Background and Assumptions

Vegetation, especially grass sod, has traditionally been one of the primary means of achieving on-site erosion control, either by shielding soil to reduce rainfall impact energy or by reducing the transport capacity of flow as is typically the case in grass waterways. Vegetation represents a relatively inexpensive means of achieving these tasks. Considerable interest has been displayed recently in the use of vegetative filters, or buffer strips, between the point of sediment detachment and its entrance into waterways.

The effectiveness of vegetative filters has been demonstrated by various researchers such as Neibling and Alberts.[12] They reported that sod strips as short as 0.61 m reduced sediment discharge rates by more than 90% and that sediment discharge of particles less than 0.002 mm through 0.61-m strips was reduced by 37%. Although their results show trapping efficiencies for the smaller particles higher than any comparable tests conducted by the authors of this section, they do indicate that satisfactorily designed grass filters may indeed enable substantial reduction of sediment problems at considerable cost savings.

Traditionally, the design of such filters has been based on the user's experience. An alternative method has been developed, which utilizes a series of equations that can be used to estimate filter performance based on particle size distribution, flow rate, sediment load, filter channel slope, and filter dimensions. Data collected using both simulated and real grasses have shown a high correlation between observed and estimated values.[9] These procedures are very tentative, and much additional work needs to be done in this area to determine the usefulness of the procedures. These procedures consider all particles to be noncohesive and noncolloidal, including those clays that may exhibit an electrostatic affinity for grass or actually be colloidal. Since these attributes would increase trapping, the estimate given by the design procedures should be conservative. A graphical solution has been developed, which will be discussed in sufficient detail so that a user can determine if the procedures can be applied to his needs. A sample design is discussed that demonstrates the fundamentals of grass filter design and enables relatively inexperienced users to design filters.

The following assumptions should be considered by the user.

1. The grass must be erect and nonsubmerged. Considerable work conducted on grass waterways indicates that maximum retardation of flow occurs prior to vegetation submergence. Also, submergence may lead to the grass stems being laid over (approximately parallel to the ground surface) so that inundation by sediment may occur. This will subsequently retard or eliminate vegetation growth. Note that this assumption conflicts with grass waterway design.
2. Incoming sediment load is greater than the transport capacity of the flow through the filter. No research has been completed that shows quantitatively the conditions under which deposited sediment is eroded from a grass filter.
3. The influent sediment concentration does not appreciably alter the viscosity of the

water with the water temperature assumed to be 25°C. Viscosity was not significantly varied during verification even though the equations do contain viscosity as an input.

4. Reduction of sediment yield results only from retardation of flow. The effect of infiltration reducing the surface runoff is neglected, although Hayes et al.[7] showed that significant trapping can result from infiltration. If desired, infiltration can be easily incorporated.

Determination of Vegetation Parameters
Flow Rate Effects and Filter Geometry

Selection of a design storm is based upon criteria similar to those discussed previously for grassed waterways. The peak discharge from this storm is used as the design discharge. However, an additional functional requirement is that the vegetation must not be overtopped, or submerged, in order to retard flow. Thus, grass height and filter width are important and must be related to peak discharge. Based on observations in the laboratory, grasses such as ryegrass and fescue will probably have to be mowed during the growing season in order to maintain rigidity. Relatively low flows can lay these grasses over if uncut, whereas stiff vegetation like grain sorghum remains rigid under high flows. Height of the grass controls the maximum depth of flow and also the storage capacity of the filter. Storage is affected because the depth of sediment deposition will not exceed the media height. Filter length is usually limited because of physical restraints at the site.

Quantitative tests have not been conducted to determine maximum grass heights that can be used without bending over the grass. A reasonable maximum value based on observations seems to be about 15 cm, although this value varies depending on the exact hydraulic variables of the situation. A design grass height can also be estimated based on mowed heights and grass species if the noted problem areas are avoided. For example, if the grass is to be mowed to a height of 10 cm, design height should be limited to 10 cm or less so that the grass will not become submerged.

In order to calculate flow depth within the filter, grass spacing, retardance coefficient, and stem diameter are used to calculate flow velocity. The modified version of Manning's formula with metric units as used in the model development is[22]

$$V_m = \frac{1}{n} R^{2/3} S^{1/2} \tag{6}$$

where V_m is mean flow velocity, n is Manning's roughness coefficient modified for grass filters, R_s is spacing hydraulic radius, and S_c is slope of the energy gradeline.

The spacing hydraulic radius is defined as

$$R_s = \frac{S_s d_f}{2d_f + S_s} \tag{7}$$

where d_f is the depth of flow and S_s is the spacing between filter media elements. The spacing can be estimated by taking the square root of the soil surface area per plant. A rough estimate of spacing, based on visual comparison, may be made using Table 5.[7]

Table 5 also shows values of the roughness coefficient n that can be used. Care must be taken not to use values of n developed for grassed waterways since these values are *not* interchangeable with the n values used in these grass filter relations, although the equations are similar. The values in Table 5 are rough estimates only and should be treated as such. Much additional research is needed in this area in order to elucidate these values for additional vegetation species.

Table 5
SPACINGS USED FOR GRASS FILTER
DESIGN

Grass condition	Spacing (cm)	Roughness coefficient (sec/cm 1/3)
Fair stand	2.0	0.008
Good stand	1.7	0.012
Dense stand	1.4	0.016

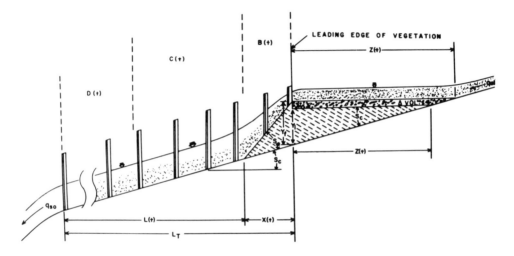

FIGURE 7.   Schematic of the sediment deposition process with a rigid medium and deposition upstream of the filter (prior to time t.).

Required Basic Relationships

Observations of numerous tests using laboratory flumes have indicated several tendencies that typify grass filters. These concepts have been discussed previously by Barfield et al.[1] and will only be briefly discussed here using the conceptual flow schematic shown in Figure 7. Only the case of steady-state flow will be considered.

In order to understand the mechanisms illustrated in Figure 7, we will first look at zone $D(t)$. In this zone, insufficient material has reached the bed for bedload to occur so that all material reaching the bed is assumed to be trapped. By resorting to probabilistic reasoning, Tollner et al.[22] showed that the fraction of sediment trapped in zone D, $(q_{sd} - q_{so})/q_{sd}$, could be predicted from the effective filter length, $L(t)$, flow depth, d_f, flow velocity, V_m, settling velocity, V_s, and grass spacing, S_s, or:

$$\frac{q_{sd} - q_{so}}{q_{sd}} = \exp\left\{ -1.05 \times 10^{-3} \left[\frac{V_m R_s}{\nu}\right]^{0.82} \left[\frac{V_s L(t)}{V_m d_f}\right]^{-0.91} \right\} \tag{8}$$

By assuming an analogy between flow in a filter medium and that in a rectangular channel of S_s and depth d_f, Manning's equation was modified to calculate flow velocity and depth of flow. Using these values in Equation 6, the value of the outflow sediment load q_{so} can be calculated for a given fall velocity if the effective filter length and sediment load into zone $C(t)$, q_{sd}, shown in Figure 7 is known.

In order to calculate a value for q_{sd}, sediment transport relationships were needed for flow within the filter. Tollner et al.[23] modified Einstein's and Graf's transport and shear parameters

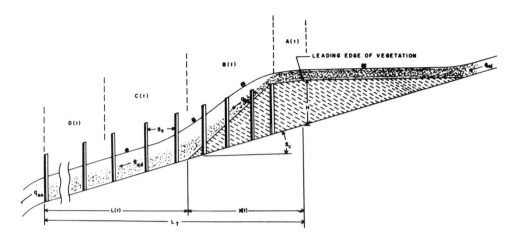

FIGURE 8.   Schematic of the sediment deposition process with a rigid medium and deposition upstream of the filter (after time t.).

to account for the presence of vegetation and also calibrated these equations for flow in a filter medium. Either of these modified equations can be used to predict a value for q_{sd}. We have chosen to use Einstein's equations because of their widespread use.

Calculation of the effective filter length, L(t), requires a prediction of how the deposition profile advances with time. When a sediment-laden flow is introduced into vegetation, the flow depth increases and velocity decreases because of the retarding effect of the vegetative elements. The decreased flow velocity results in decreased sediment transport capacity. If the sediment transport capacity is then sufficiently less than the input sediment load, deposited particles will form a triangular-shaped wedge as shown in Figure 7. If this continues with constant inflow conditions, the deposition wedge will approach the same height as the vegetation at time t. At this height, the wedge will flatten out but continue to advance downstream at approximately the same angle compared to horizontal as shown in zones A(t) and B(t) of Figure 8. By performing a mass balance using the slope of the deposition wedge, S_e, the advance distance can be computed using either Equation 7 or 8 in Table 6, depending on the time. The value of S_e can be calculated from the modified Einstein and Manning equations mentioned previously. A summary of the flow and transport equations is given in Table 6 for zones B(t), C(t), and D(t) even though they are not discussed in detail here. Most of them are incorporated into the graphical solution. While these equations are complex and require an iterative approach for solution, they can be solved relatively easily using the graphical solution which is described in a subsequent section.

The aforementioned analyses consider the transport capacity within the vegetative medium, but do not consider the possible effect upstream trapping may have on sediment load. Work reported by Neibling and Alberts[12] indicates that there may be significant trapping in this area. To incorporate this component into the design procedures, Hayes et al.[8] utilized a mass balance to calculate the sediment trapped in this region (see Figure 7). The depth of deposition Y(t) has previously been estimated by:

$$Y(t) = \frac{\overline{2\,fq_s}}{\gamma_{sb}} S_c^{1/2} \tag{9}$$

If Equation 9 is solved at two different times, t_1 and t_2 less than t., while assuming that the deposition wedge profile extends upstream horizontally until it intersects the channel bottom, the sediment load, q_{si}, reaching the filter can be calculated by:

Table 6
SUMMARY OF THE FLOW AND TRANSPORT EQUATIONS FOR STEADY-STATE, HOMOGENEOUS SEDIMENT BASED ON UNIT WIDTH OF FILTER

Zone D(t)
  Sediment trapping

$$\frac{q_{sd} - q_{so}}{q_{sd}} = \exp\left\{-1.05 \times 10^{-3}\left[\frac{V_m R_s}{\nu}\right]^{0.82}\left[\frac{V_s L(t)}{V_m d_f}\right]^{-0.91}\right\} \tag{1}$$

  Flow

$$V_m = \frac{1}{n} R_s^{2/3} S_c^{1/2} \tag{2}$$

$$R_s = \frac{S_s d_f}{2d_f + S_s} \tag{3}$$

$$q_w = V_m d_f \tag{4}$$

Zone C(t)
  Flow
    Same as Zone D(t)
  Sediment transport

$$\frac{\gamma_s - \gamma}{\gamma}\left[\frac{d_p}{S_c R_s}\right] = 1.08\left[\frac{q_{sd}}{\gamma_s\sqrt{\left(\frac{\gamma_s \gamma}{\gamma}\right) g d_p^3}}\right]^{-0.28} \tag{5}$$

Zone B(t)
  Time of advance to media height

$$t_* = \frac{H^2 \gamma_{sb}}{2 f q_{si} S_e} \tag{6}$$

  Rate of advance

$$X(t) = \left(\frac{\overline{2f \, q_s} t}{\gamma_{sb} \, S_e}\right)^{1/2}, \, t < t_* \tag{7}$$

$$X(t) = \frac{\overline{f \, q_s}(t - t_*)}{H_{\gamma sb}} + X(t_*), \, t > t_* \tag{8}$$

$$f = \frac{q_s - q_{sd}}{q_s} \tag{9}$$

$$S_e = S_{et} - S_c \tag{10}$$

  Sediment transport

$$q_{ss} = \frac{q_s + q_{sd}}{2} \tag{11}$$

$$\frac{\gamma_s - \gamma}{\gamma}\left[\frac{d_p}{R_{ss} S_{et}}\right] = 1.08\left[\frac{q_{ss}}{\gamma_s\sqrt{\left(\frac{\gamma_s - \gamma}{\gamma}\right) g d_p^3}}\right]^{-0.28} \tag{12}$$

Table 6 (continued)
SUMMARY OF THE FLOW AND TRANSPORT EQUATIONS FOR STEADY-STATE, HOMOGENEOUS SEDIMENT BASED ON UNIT WIDTH OF FILTER

$$R_{ss} = \frac{S_s \, d_{fs}}{2d_{fs} + S_s} \tag{13}$$

$$q_w = V_{ss} \, d_{fs} \tag{14}$$

$$V_{ss} = \frac{1}{n} R_{ss}^{2/3} S_{et}^{1/2} \tag{15}$$

$$q_{si} = q_s - \frac{(Y_2^2 - Y_1^2) \, \gamma_{sb}}{2 \, S_e(t_2 - t_1)} \tag{10}$$

where q_s is the initial sediment load. Equation 10 is the model for predicting the effects of upstream deposition on the sediment load.

Influent Sediment Relationships

Exact quantitative relationships are presently unavailable for specification of design sediment loads and sediment properties. Typically, the peak flow from a certain design storm is used as the design discharge. The peak flow, Q_p, and associated runoff volume, Q, can be calculated by a number of methods depending on the preference of the user. Sediment concentrations are known to change throughout a storm event. A satisfactory methodology for describing this process based on readily available parameters is not available. Thus, concentration is often assumed to be independent of discharge, and average concentration is assumed throughout the storm event. The mass of sediment, M, displaced during the design storm can be estimated using either the Universal Soil Loss Equation (USLE) with an appropriate delivery ratio or other sediment yield relations. Average concentration, C_{av}, can be calculated by

$$C_{av} = \frac{M}{Q_v} \tag{11}$$

where M is the sediment yield and Q_v is runoff volume resulting from the design storm. The design discharge per unit width, q_w, is given by

$$q_w = \frac{Q_p}{W} \tag{12}$$

where Q_p is the peak flow for the design storm and W is the width of the filter in the direction perpendicular to the flow. The design sediment discharge per unit width, q_s, is now calculated by

$$q_s = q_w \cdot C_a \tag{13}$$

where q_w is peak flow per unit width and C_a is average concentration.

Since inflow sediment is almost always composed of a wide range in sizes, a particle size distribution curve similar to the one shown in Figure 9 is usually determined prior to calculating filter efficiency. In order to simplify this distribution for hand calculations, a

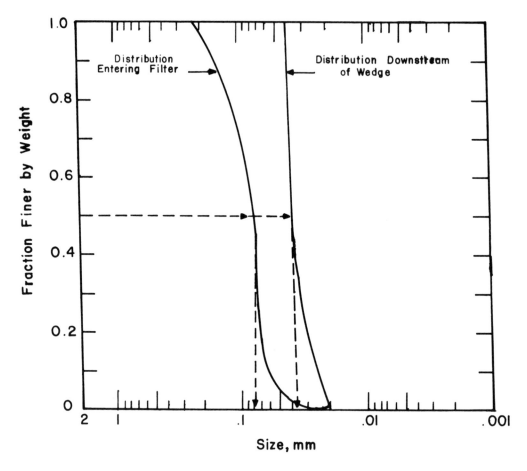

FIGURE 9.   Example of an inflow particle size distribution curve indicating how the size distribution can be modified to account for fraction trapped.

single size should be selected as the representative particle diameter. (Note that in this discussion, particle refers to either aggregated particles or primary particles depending on how they enter the filter.) The diameter selected is usually the mean diameter by weight, d_{50}, which corresponds to the diameter where the particle size distribution curve intersects the 50% finer line in Figure 9. A standardized procedure for obtaining a size distribution to be used in design of sediment controls has not been established to date. Several alternative methods are available for obtaining a sample to be tested for size distribution. Additional work is needed in order to define the "best" procedure. The user should therefore consider the possible consequences if an erroneous particle diameter is used in design calculations.

A simple procedure is to analyze a sample of runoff from the drainage area at the point where the flow will eventually enter the filter. Runoff from an adjacent watershed with similar runoff and erosive characteristics may also be used if necessary. Analysis using pipette withdrawals at selected time intervals based on settling velocities is the most commonly accepted method of obtaining the size distribution curve from a field sample. Again it should be noted that the sample analyzed is composed of both aggregated and primary particles so that the material should not be dispersed prior to pipetting. Possible errors that may occur during evaluation of sediment size distributions are discussed by Meyer and Scott.[11]

In the absence of established procedures to estimate the size distribution of eroded sediment, Barfield et al.[2] suggested taking a sample of surface soil and using a simple rainfall

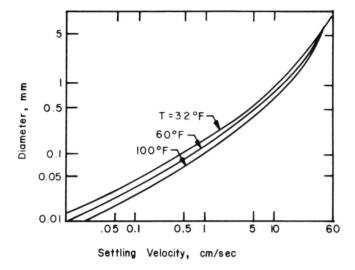

FIGURE 10.   Chart of sedimentation diameter vs. fall velocity in water assuming a specific gravity of 2.65. (From Haan, C. T. and Barfield, B. J., Hydrology and Sedimentology of Surface Mined Lands, Office of Continuing Education and Extension, College of Engineering, University of Kentucky, Lexington, 1978. With permission.)

simulator system composed of a single Spraying Systems Vee Jet 80150 nozzle. The nozzle is located 3 m above a small flat-bottomed container which discharges onto a nest of sieves in a bucket. After wet sieving, the fine material in the bucket is pipetted. The system is operated until the volume of rainfall equals that of the desired storm. The authors note that *this procedure has not been experimentally verified*, but is better than a guess.

A third procedure that has been utilized is to obtain a sample of the surface soil in the watershed. This sample can then be pipetted for size distribution (without dispersion). Difficulties with this procedure arise because the resulting distribution will typically be biased by larger particles that will tend to be easily trapped prior to reaching the sediment control structure. If the diameter used for design calculations is too large, the predicted fraction trapped will also result in a high value.

After a representative diameter is estimated, the particle settling velocity, V_s, can be found using Figure 10 at the appropriate water temperature.

Selection of Vegetation, Establishment, and Maintenance

Preparation is a prime factor in successful installation of a vegetative filter. Even if the filter is suitably designed, poor installation or maintenance can render it useless. The filter installation should always take place during optimum weather conditions in order to avoid periods when rainfall intensities are expected to be very high. Otherwise, loosely packed soil may be eroded easily. Erosion during establishment can be controlled by a number of methods including mulches, chemical stabilizers, and diversions. The four major steps that are involved in the installation of a grass filter include: (1) site selection, (2) vegetation selection, (3) establishment of the filter, and (4) maintenance. Each of these phases will be discussed in more detail.

Site Selection

In most field situations, suitable sites will be limited in number because of the prevailing drainage pattern in the area. Movement of large volumes of soil material is to be avoided. Thus, a general topographic survey of potential sites should be initiated early in the design stages. Also, diversion of water during establishment may be desirable. The site selected must be such that inflow can be distributed evenly across the upslope end of the filter to prevent channelization within the channel. In order to maintain this distribution, the filter must be nearly flat in the direction perpendicular to the flow. In some instances where the entering flow is concentrated, baffles or other energy-retarding structures are required up-stream from the grass. Filter length in the flow direction is usually curbed by physical restraints such as ditches, waterways, or fields.

Field reconnaissance should also include a soil test. Results from this test will usually give insight into which species of vegetation might be suited to the site. For example, some species grow well in acid soils, while others grow best in alkaline soils. Subsoils should also be checked since in some instances subsoil may have substantially different pH, or some other property, compared to topsoil. This condition may lead to inhibited root growth in the subsoil if the vegetation is selected based only on topsoil conditions. Results from the soil test will also include specific fertilizer recommendations for both planting and later stages of growth. Equipment and complete instructions for obtaining soil tests can usually be obtained by contacting the local county extension agent.

Vegetation Selection

In addition to selecting a vegetation that is tolerant of chemical stresses such as the aforementioned ones, the vegetation should be capable of enduring stresses caused by climatic conditions. These conditions may include hot and cold temperatures or severly dry to wet conditions. The vegetation should be selected so that it will be actively growing during periods of greatest erosion potential instead of dormant. As an example, cool season grasses are preferred in areas that receive heavy spring rains with little cover on the area upslope from the filter. The grass should also provide quick establishment to permit early protection of the filter area itself by reducing chances for rills to form. Finally, the vegetation must consist of strong, erect stems that will not flatten and allow the filter to act like a grass waterway. In order to fulfill all of these requirements, two or more species may be mixed.

Establishment of the Filter

Establishment of vegetation necessitates preparation of a good seedbed. The soil should be conditioned by discing, harrowing, or similar means. Care must be taken to maintain a level seedbed in the direction perpendicular to future flows. Fertilizer and lime can also be applied and incorporated at this stage as called for by the soil test recommendations. Seed can be broadcast using a cyclone spreader, preferably after dividing the seed into several portions so that the seed can be distributed from a number of directions. This procedure will assist in obtaining uniform seeding rates throughout the filter. Seeding rates will be controlled by the vegetation species and by the number of plants per unit area that are required to give the desired spacing. Although most grasses tend to reach an ultimate number of stems per unit area, high seeding rates are often desirable in order to quickly reach this quantity. Thus, a typical seeding rate might be on the order of 125 kg/ha (110 lb/acre). Such a rate is approximately five times the normal seeding rate for pastures or one half that of lawns. During and after seeding, attention should be placed on avoidance of field patterns or wheel tracks that may concentrate flow and eventually cause rills or gullies. The seed can be incorporated by light discing or harrowing to increase germination and reduce likelihood of seed being displaced by rainfall. For large areas, hydroseeding is advantageous. Mulches such as wheat straw and/or chemical stabilizers can also be used to advantage in order to reduce erosion potential during establishment and prevent drying of the soil. Packing of the filter area with a roller can also reduce erosion while enhancing seed germination.

Table 7
INPUTS FOR GRAPHICAL
SOLUTION EXAMPLE

Filter length (L)	15 m
Grass height (H)	15 cm
Spacing (S_s)	1.6 cm
Channel slope (S_c)	0.02
Manning's n (n)	0.012
Average particle size (d_p)	0.0074 cm
Particle settling velocity (V_s)	0.38 cm/sec
Sediment load (q_s)	7.50 g/sec-cm
Water flowrate (q_w)	50.0 cm^3/sec-cm
Sediment bulk density (γ_{sb})	1.5 g/cm^3

Maintenance

Although maintenance requirements for grass filters are relatively minor, they must not be overlooked or filter failure may result. Routine management necessitates occasional mowing of the grass so that it will not easily lay over under high flow conditions. Mowing should occur at least once per month during the active growth periods of the vegetation. Care should be taken to avoid leaving large quantities of grass clippings on the surface, which may limit growth. A mowing pattern should be used that does not leave wheel-packed grass in the direction of pending flow. During establishment and under high stress conditions later, irrigation may be beneficial. Occasional insect and/or weed control may be desirable, although significant problems in these areas are unlikely.

Graphical Solution Procedure

In order to increase the usefulness of the grass filter design model, a graphical solution has been developed that can be utilized to rapidly estimate the efficiency of a proposed filter for steady-state conditions. The precision of the graphical solution is limited primarily by the user's ability to read the scales. The solution technique is based upon the same equations that were presented previously including Manning's equation, equation of continuity, and Einstein's total transport function, each modified for use in a grass filter. Additional information explaining this development may be found in Hayes et al.[9]

Example Problem No. 2

The following example illustrates the solution procedure. In order to apply the graphical solution technique, input information concerning the expected filter conditions is necessary. The information is shown in Table 7 for the example. The filter length and channel slope are based on a field survey. The spacing, Manning's n, and grass height are assumed (see Table 5). The particle size distribution curve shown in Figure 9 is assumed so that the inflow particle size is d_{50} and its settling velocity is found from Figure 10. Sediment load and flow rate were derived using the design storm on a per unit width basis (width assumed). With this information, four terms that need to be calculated are

$$q_w n/S_c^{1/2}, \quad S_c^{3.57}/d_p^{2.07}, \quad q_w n/0.622 \, d_p^{0.29}, \quad \text{and} \quad q_w n$$

The following steps may then be followed to calculate the trapping in the deposition wedge:

1. $q_w n/S_c^{1/2} = 4.24$ at the start position in Figure 11.
2. The intersection point of the constant $q_w n/S_c^{1/2}$ line with the constant spacing line

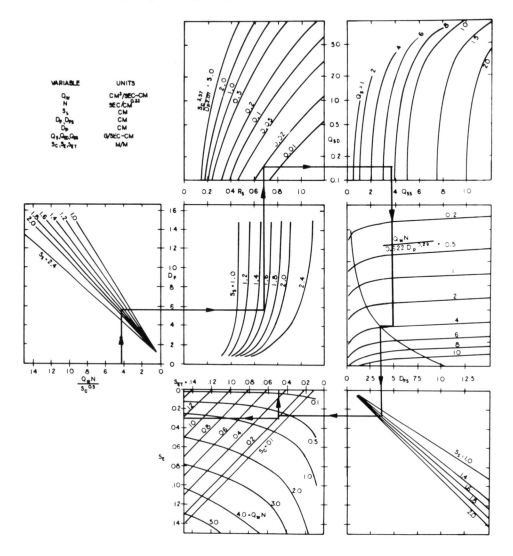

FIGURE 11. Nomographs used to calculate depth of flow and equilibrium slope.

(e.g., S_s = 1.6) gives the value of the depth of flow (e.g., d_f = 5.4). The flow depth should be checked to insure that flow depth does not exceed the height of the grass.

3. Extending the constant depth line to intersect with the spacing line (e.g., S_s = 1.6) yields a value of the spacing hydraulic radius (e.g., R_s = 0.71).

4. Extension of the constant spacing hydraulic radius line to the constant $S_c^{3.57}/d_p^{2.07}$ = 0.022 line gives the sediment load that can be transported downstream of the sediment wedge (e.g., q_{sd} = 0.19).

5. The intersection of the constant q_{sd} line with the constant inflow sediment load line (q_s = 7.5) gives the average sediment load on the deposition wedge (e.g., q_{ss} = 3.85).

6. Continuation of the average load on the wedge line to the constant $q_w n/0.622\, d_p^{0.29}$ (e.g., $q_w n/0.622\, d_p^{0.29}$ = 4.00), then turning horizontally to the turn line before extending the line yields the value of the flow depth on the deposition wedge (e.g., d_{fs} = 3.6).

7. Extend the constant d_{fs} line down to its intersection with the constant spacing line (e.g., S_s = 1.6).

8.   Continuing across to the constant $q_w n$ line (e.g., $q_w n = 0.60$), the absolute slope of the deposition wedge can be found by extending the line vertically and reading the S_{et} axis (e.g., $S_{et} = 0.049$) or one can go vertically to the constant channel slope line (e.g., $S_c = 0.02$) and then go across to read the equilibrium slope (e.g., $S_e = 0.029$).
9.   The next step is to calculate the fraction trapped on the deposition wedge using the equation

$$f = \frac{q_s - q_{sd}}{q_s} = 0.975$$

10.   At this point, everything needed is known, and the time at which the deposition wedge reaches the media height can be calculated from Equation 6 in Table 6:

$$t_* = \frac{H_{sb}^2}{2fq_s S_e} = 800 \text{ sec}$$

11.   By picking two times (t_1 and t_2) and using Equations 9 and 10, an estimate of what effect ponding upstream from the deposition wedge will have on inflow sediment can be obtained. If q_{si} is not approximately equal to q_s, increased accuracy can be obtained by repeating the analysis letting q_{si} take the place of q_s.
12.   The deposition wedge advance distance can now be calculated using Equation 7 in Table 6 if $t < t_*$ or using Equation 8 in Table 6 if $t > t_*$.

Calculation of the trapping efficiency downstream of the sediment wedge can be accomplished by using the known inputs and the values of spacing hydraulic radius (R_s), sediment load downstream from the wedge (q_{sd}), and flow depth (d_f) from the previous analysis. This analysis is based on Equation 8. If a particle size distribution is available, the fraction of inflow sediment trapped on the deposition wedge should be subtracted and a new representative particle size chosen based on the remaining size distribution. This size and its respective settling velocity should be used to estimate the downstream trapping as shown in zone D(t) of Figure 7.

The effective length of the filter can be calculated by subtracting the wedge advance distance from the original filter length. For this example, the same conditions as for previous analysis will be assumed except for the following:

1.   Assume that the advance distance, X(t), is 5 m so that the effective length (L) equals 10 m.
2.   Assume that the new representative particle diameter is 0.0037 cm and its settling velocity is 0.10 cm/sec as shown in Figures 9 and 10 for the distribution downstream from the wedge.

With this information, the following steps can be followed using Figure 12:

1.   Follow the line of constant channel slope (e.g., $S_c = 0.02$) to the line of constant spacing hydraulic radius found in the previous analysis (e.g., $R_s = 0.71$).
2.   Extend the line across to the constant flow depth line found in the previous analysis (e.g., $d_f = 5.4$).
3.   Turn vertically until the constant flow depth line intersects with the constant Manning's n line (e.g., $n = 0.012$).
4.   Continue by moving horizontally across to the line of constant settling velocity (e.g., $V_s = 0.10$).

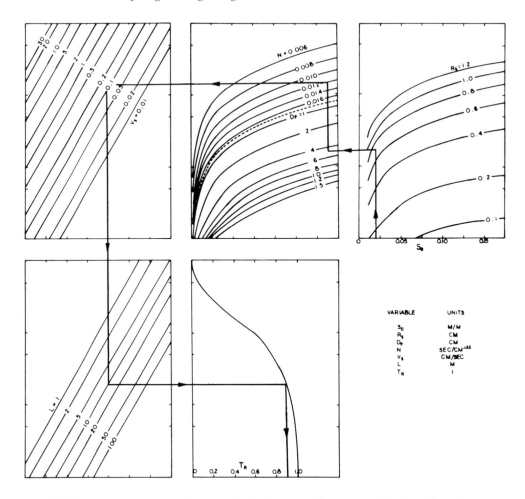

FIGURE 12.   Nomographs used to determine fraction trapped downstream of the deposition wedge.

5.   Turn vertically and go down to intersect the line of constant filter length (e.g., L = 10).

6.   The trapping efficiency for the sediment load downstream of the deposition wedge can then be obtained by turning the line horizontally to the trapping function line so that $T_r = 0.88$.

7.   The outflow sediment load can be calculated by multiplying one minus the trapping efficiency downstream times the sediment load coming into the downstream portion (e.g., $q_{sd} = 0.19$) so that:

$$q_{so} = (1 - T_r)q_{sd} = 0.023$$

8.   The total trapping efficiency can be obtained by exchanging q_{so} for q_{sd} in Equation 9 in Table 6 and substituting:

$$f = \frac{q_s - q_{so}}{q_s} = 0.997$$

9.   The outflow concentration can also be calculated by multiplying the fraction remaining, (1-f), times the inflow concentration or:

$$C_o = (1 - f) C_i$$

where c_o is outflow concentration, f is total fraction trapped, and C_i is inflow concentration.

In instances where filter length is relatively unlimited, the procedure can be solved by trial and error to obtain a specific fraction trapped. Alternatively, the total fraction trapped can be utilized to find the necessary filter length by reversing the procedure in the second nomograph to find the length of filter needed in the downstream portion. If this is done, one *must* remember to add the length corresponding to the advance distance or insufficient storage capacity will be contained within the filter.

REFERENCES

1. **Barfield, B. J., Hayes, J. C., and Barnhisel, R. I.,** The Use of Grass Filters for Sediment Control in Strip Mine Drainage, Vol. 2, IMMR39-RRR4-78, Institute for Mining and Minerals Research, University of Kentucky, Lexington, 1978.
2. **Barfield, B. J., More, I. D., and Williams, R. G.,** Prediction of Sediment Yield from Surface Mines Watersheds, Symp. Surface Mining Hydrology, Sedimentology, and Reclamation, University of Kentucky, Lexington, 1979.
3. **Fenzl, R. N. and Davis, J. R.,** Hydraulic resistance relationships for surface flows in vegetated channels, *Trans. ASAE,* 7(1), 46, 1972.
4. **Gwinn, W. R. and Ree, W. O.,** Maintenance effects on the hydraulic properties of a vegetation lined channel, *Trans. ASAE,* 23(3), 636, 1980.
5. **Haan, C. T.,** *Statistical Methods in Hydrology,* Iowa State University Press, Ames, 1977.
6. **Haan, C. T. and Barfield, B. J.,** Hydrology and Sedimentology of Surface Mined Lands, Office of Continuing Education and Extension, College of Engineering, University of Kentucky, Lexington, 1978.
7. **Hayes, J. C., Barfield, B. J., and Barnhisel, R. I.,** Evaluation of Grass Characteristics Related to Sediment Filtration, Paper No. 78-2513, presented at Winter Meeting of the American Society of Agricultural Engineers, 1978.
8. **Hayes, J. C., Barfield, B. J., and Barnhisel, R. I.,** Evaluation of Vegetal Filtration for Reducing Sediment in Surface Mine Runoff, Symp. Surface Mining Hydrology, Sedimentology, and Hydrology, University of Kentucky, Lexington, 1979.
9. **Hayes, J. C., Barfield, B. J., and Barnhisel, R. I.,** The Use of Grass Filters for Sediment Control in Strip Mine Drainage, Vol. 3, Institute for Mining and Minerals Research, University of Kentucky, Lexington, in press.
10. **Larson, C. L. and Manbeck, D. M.,** Improved procedures in grassed waterway design, *Agric. Eng.,* 41, 694, 1960.
11. **Meyer, L. D. and Scott, S. H.,** Possible Errors During Field Evaluations of Sediment Size Distributions, Paper No. 81-2044, presented at Summer Meeting of the American Society of Agricultural Engineers, 1981.
12. **Neibling, W. H. and Alberts, E. E.,** Composition and Yield of Soil Particles Transported through Sod Strips, Paper No. 79-2065, presented at Summer Meeting of the American Society of Agricultural Engineers, 1979.
13. **Palmer, R. S.,** River-terrace diking for fully control, *Trans. ASAE,* 6(4), 325, 1963.
14. **Ree, W. O.,** Hydraulic characteristics of vegetation for vegetated waterways, *Agric. Eng.,* 30, 184, 1949.
15. **Ree, W. O.,** Retardation coefficients for row crops in diversion terraces, *Trans. ASAE,* 1(1), 78, 1958.
16. **Ree, W. O.,** Effect of Seepage Flow on Reed Canarygrass and Its Ability to Protect Waterways, ARS-S-154, ARS, U.S. Department of Agriculture, Washington, D.C., 1976.
17. **Ree, W. O.,** Performance Characteristics of a Grassed Waterway Transition, ARS-S-158, ARS, U.S. Department of Agriculture, Washington, D.C., 1977.
18. **Ree, W. O. and Palmer, V. J.,** Flow of Water in Channels Protected by Vegetative Linings, Tech. Bull. 967, U.S. Department of Agriculture, Washington, D.C., 1949.
19. **Schwab, G. O., Frevert, R. K., Edminster, T. W., and Barnes, K. K.,** *Soil and Water Conservation Engineering,* John Wiley & Sons, New York, 1966.

20. Soil Conservation Service, Handbook of Channel Design for Soil and Water Conservation Practices, U.S. Department of Agriculture, Washington, D.C., 1974.

21. **Temple, D. M.,** Tractive force design of vegetated channels, *Trans. ASAE,* 23(4), 884, 1980.

22. **Tollner, E. W., Barfield, B. J., Haan, C. T., and Kas, T. Y.,** Suspended sediment filtration capacity of simulated vegetation, *Trans. ASAE,* 19(4), 678, 1976.

23. **Tollner, E. W., Hayes, J., and Barfield, B. J.,** The Use of Grass Filters for Sediment Control in Strip Mine Drainage, Vol. 1, IMMR35-RRR2-78, Institute for Mining and Minerals Research, University of Kentucky, Lexington, 1978.

24. Soil Conservation Service, Handbook of Channel Design for Soil and Water Conservation, SI 1966, U.S. Department of Agriculture, Washington, D.C., 1966.

WIND EROSION CONTROL

Leon Lyles

DEFINITION OF PROBLEM

Wind erosion is caused by strong, turbulent winds over unprotected soil surfaces that are smooth, bare, loose, dry, and finely granulated. The extremely complex process involves initiation, transport, sorting, abrasion, avalanching, and finally soil-particle deposition. Wind erosion affects environmental quality in several ways. Dust obscures visibility, pollutes the air, causes automobile accidents, and fouls machinery and electrical apparatus. Blowing soils abrade and kill plants, reducing quantity and quality of food supplies. Uncontrolled wind erosion deposits soil that buries irrigation ditches, fences, and roads. The selective or total removal of the active soil components by wind reduces the productivity of the land.

In the U.S., about 66 million ha of land — 22.5 million ha of cropland, 43 million ha of rangeland, and 0.5 million ha of forest land — are susceptible to wind erosion at amounts greater than 11 ton/ha/year. The greatest potential, however, centers in the ten Great Plains states that contain 80% of the susceptible cropland and 42% of the susceptible rangeland. In these states, about 2 million ha (average) of land are moderately to severely damaged by wind erosion each year. Outside the U.S., agricultural areas susceptible to wind erosion include much of North Africa and the Near East, parts of southern and eastern Asia, Australia, and southern South America.

PARTICLE MOTION

Soil-particle movement by wind is generally described in three distinct modes: suspension, saltation, and surface creep. Particles in suspension can range in size from 2 to 100 μm in diameter, with a mass median diameter of about 50 μm in an eroding field.[1,2] In long-distance transport, however, particles <20 μm in diameter predominate.[3] Some suspension-size particles are present in soil, but many are created by abrasive breakdown during erosion. Depending on soil texture, 3 to 38% of the eroding soil could be carried in suspension.[4] Generally, the vertical flux is <10% of the horizontal.[3,5] Suspended particles are best reduced by good wind-erosion control practices; for example, tall vegetation can sometimes be used to trap particles already suspended.[6,7]

The characteristics of saltation (jumping) particles in wind have been described by several scientists.[4,8-11] In saltation, individual particles lift off the surface (eject) at angles of 50 to 90° and follow distinctive trajectories under the influence of air resistance and gravity.[4,11] Such particles, 100 to 500 μm in diameter (too large to be suspended by the flow), return to the surface at impact angles of 6 to 14° from the horizontal either to rebound or to embed themselves and initiate movement of other particles. Saltation constitutes roughly 50 to 80% of the total transport.

Mineral-soil (sand) particles or aggregates 500 to 1000 μm in diameter (too large to leave the surface in ordinary erosive winds) are set in motion by the impacts of saltating particles. In high winds, the whole sand/soil surface appears to be creeping slowly forward at speeds ≤2.5 cm/sec — pushed and rolled (driven) by the saltation flow. Reportedly, surface creep constitutes 7 to 25% of the saltation flow.[4,9,12]

Table 1

**SOIL ERODIBILITY I FOR SOILS WITH DIFFERENT
PERCENTAGES OF NONERODIBLE FRACTIONS AS
DETERMINED BY STANDARD DRY SIEVING**

Percentage of Dry Soil Fractions > 0.84 mm

Tens	Units (metric tons/ha/year)									
	0	1	2	3	4	5	6	7	8	9
0	—	695	560	493	437	404	381	359	336	314
10	300	294	287	280	271	262	253	244	238	228
20	220	213	206	202	197	193	186	182	177	170
30	166	161	159	155	150	146	141	139	134	130
40	126	121	117	114	112	108	105	101	96	92
50	85	80	75	70	65	61	28	54	52	49
60	47	45	43	40	38	36	36	34	31	29
70	27	25	22	18	16	13	9	7	7	4
80	4	—	—	—	—	—	—	—	—	—

From Woodruff, N. P. and Siddoway, F. H., *Soil Sci. Soc. Am. Proc.*, 29, 602,
1965. With permission.

PRINCIPLES AND CONTROL

Wind Erosion Equation

To control wind erosion, (1) reduce wind forces on erodible particles or (2) create particles resistant to wind forces. From our knowledge of erosion processes and mechanics, four specific principles of wind-erosion control have been identified:[13] (1) establish and maintain vegetation or vegetative residues, (2) produce or bring to the soil surface nonerodible aggregates or clods, (3) reduce field width along prevailing wind erosion direction, and (4) roughen the land surface. Control practices for applying these principles vary from place to place and may change over time, along with cropping and management systems. These four principles plus a factor for climate were used to develop a wind-erosion equation that predicts potential annual erosion rates:[14]

$$E = f(I, K, C, L, V) \tag{1}$$

where E is the potential annual soil loss rate, I is the soil erodibility, K is the soil ridge-roughness factor, C is the climatic factor, L is the unsheltered median travel distance of the wind across a field, and V is the equivalent vegetative cover.

The soil-erodibility factor (I) is defined as the potential soil loss (in metric tons per hectare per year) from a wide, unsheltered, isolated field with a bare, smooth, noncrusted surface based on climatic conditions near Garden City, Kansas. It is related to surface-soil aggregates >0.84 mm in diameter and may be determined by dry sieving and using Table 1.

Soil ridge-roughness factor (K) is a measure of the effects of ridges on erosion amounts relative to a smooth surface. It is determined from height-spacing measurements of ridges caused by tillage implements and by using the following equations:

$$K = 0.773 - 0.153 \ln K_r; \ 0 < K_r \leq 8.9 \text{ cm} \tag{2}$$

$$K = 0.336 e^{0.0324 K_r}; \ 8.9 < K_r \leq 25.4 \text{ cm} \tag{3}$$

$$K_r = 4h^2/s \tag{4}$$

where h is average ridge height in centimeters, and s is average ridge spacing in centimeters. The factor K is limited to values ≤ 1.0 (if $K_r = 0$; $K = 1.0$).

The local wind-erosion climatic factor (C) is used to characterize the erosive potential of climate (windspeed and surface-soil moisture) at a particular location relative to Garden City, Kansas, which has an annual C value of 100% based on long-term climatic data.[15] Equations for computing the C factor are

$$C = 386 \frac{\overline{u}_z^3}{\left[\sum_{i=1}^{i=12} 10(P - E)_i\right]^2} \tag{5}$$

where $\overline{u}_z$ is average annual windspeed in meters per second at a standard height z of 9.1 m, i represents months, and

$$10(P - E) = 115\left(\frac{P/2.54}{1.8T + 22}\right)^{10/9}; P \geq 1.27 \text{ cm}$$

$$T \geq -1.7 \,^{\circ}\text{C} \tag{6}$$

where P is normal monthly precipitation in centimeters and T is normal monthly temperature in degrees Celsius. In Equation 6, if $P < 1.27$ cm, use 1.27; if $T < -1.7°C$, use -1.7. The annual climatic factors for the U.S. are shown in Figures 1 and 2.

From the width factor (L) in the equation, we recognize that the rate of soil flow increases with distance downwind across an eroding field, until it reaches the transport capacity of a given wind, and that winds have a prevailing direction and a preponderance in the prevailing direction.[16,17] Prevailing wind-erosion direction and preponderance values are given in Table 2 for 212 locations in the U.S. These data are used for selecting direction factors (Table 3), which are multiplied by field width (i.e., width most nearly parallel to the prevailing wind-erosion direction) to obtain L.

The equivalent vegetative-cover factor (V) is used to consider the quantity, kind, and orientation of vegetation or vegetative residues on erosion amounts. All vegetative materials, before being used to determine V, must be converted to an equivalent small-grain standard — defined as 25.4-cm lengths of dry, small-grain stalks lying flat on the soil surface (in rows perpendicular to wind direction and spaced 25.4 cm apart), with stalks parallel to the wind direction.[18] Data are presented in Figure 3 and Table 4 for converting various crops (vegetation) to their small-grain equivalent. The V factor may then be determined from the equation:

$$V = 0.2533(SG)_e^{1.363} \tag{7}$$

where $(SG)_e$ is the small-grain equivalent in kilograms per hectare. Efforts are being continued to evaluate the small-grain equivalent of feedlot manure[19] and additional crops.[20]

Relations among variables in the equation are complex, and a single equation that expresses E, as a function of the dependent variables, has not been devised. Initially, the equation was solved in a stepwise procedure, involving tables and graphs.[14,17] Now it can be solved by computer methods.[21,22]

The equation was developed primarily to estimate the potential erosion from a particular field or the field conditions necessary to reduce potential erosion to tolerable amounts. It has been used extensively in designing and evaluating control systems (practices). It can

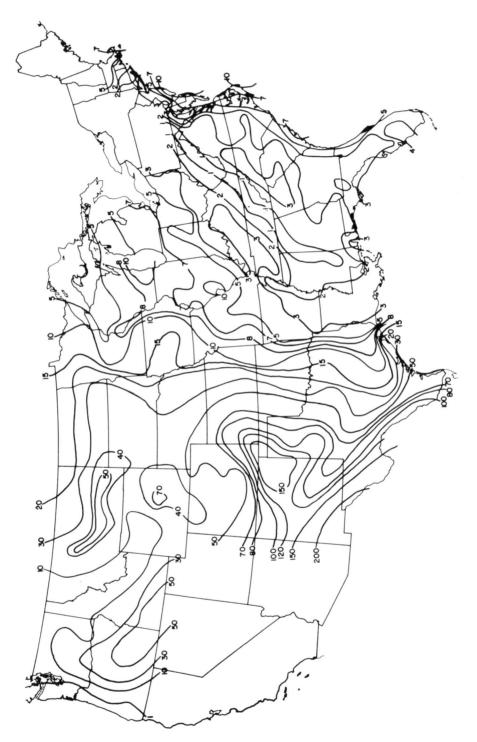

FIGURE 1. Annual climatic factor C (percent).

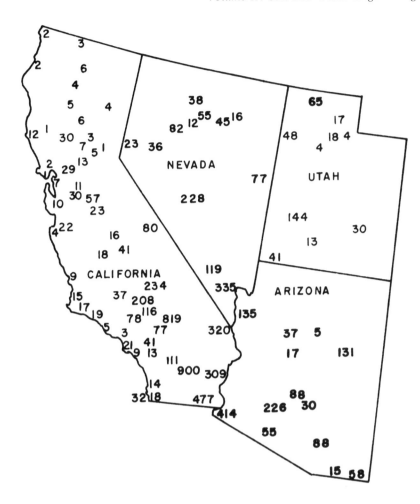

FIGURE 2.   Annual climatic factor C (percent) for Arizona, California, Nevada, and Utah.

also be used to (1) determine spacing for barriers in narrow strip-barrier systems,[23] (2) estimate fugitive dust emissions from agricultural and subdivision lands,[24,25] (3) predict horizontal soil fluxes for comparison with vertical aerosol fluxes,[26] (4) estimate effects of wind erosion on productivity,[27,28] (5) evaluate stubble requirements in field strips to trap windblown soil,[29] and (6) delineate cropland where residues might be removed without exposing the soil to wind erosion.[30]

Example Using the Wind Erosion Equation

Suppose one wanted to estimate average annual potential soil loss from a 402-m-wide field with 3200 kg/ha of standing grain sorghum stubble 30 cm tall near North Platte, Nebraska. The field has ridges spaced 100 cm apart and 5 cm high.

Sieving would show 30% of the surface soil aggregates were >0.84 mm in diameter. Table 1 shows that soil erodibility I = 166 t/ha/year. Using Equations 4 and 2, it could be determined that the soil ridge-roughness factor K is 0.77. Figure 1 shows that the climatic factor (C) is about 40% near North Platte, Nebraska.

Choosing March as the critical erosion period, refer to Table 2 to determine the prevailing wind-erosion direction at North Platte: 338° (NNW) with a preponderance of 2.1. The direction factor of about 1.36 (Table 3) times 402 m gives 547 m for L (360 to 338° = 22° angle of deviation).

Table 2

PREVAILING WIND-EROSION DIRECTION[a] (FIRST LINE) AND PREPONDERANCE OF PREVAILING WIND-EROSION FORCES IN THE PREVAILING WIND-EROSION DIRECTION (SECOND LINE)

Location	Jan	Feb	Mar	Apr	May	Jun	Jul	Aug	Sep	Oct	Nov	Dec
Alaska												
Anchorage	22	22	0	180	180	180	180	180	180	180	180	0
	2.0	3.4	2.7	2.7	3.4	3.1	2.7	3.2	3.1	3.2	3.1	3.1
Fairbanks	45	225	45	225	45	247	247	247	247	247	225	45
	2.6	2.9	2.5	1.3	1.9	2.1	2.2	2.4	2.0	2.5	1.3	3.3
Arizona												
Ajo	180	180	337	225	225	203	225	180	202	180	180	180
	1.9	2.1	1.2	1.2	1.4	1.8	1.3	1.2	1.6	1.5	1.7	1.4
Douglas	201	203	225	247	248	225	337	315	90	90	90	180
	1.3	1.6	1.5	1.4	1.5	1.6	1.4	1.4	1.5	1.8	1.0	1.0
Kingman	225	203	203	225	225	225	225	225	225	225	225	0
	2.9	1.7	2.6	2.8	2.7	3.6	2.4	2.9	1.9	1.5	2.0	1.6
Phoenix	90	293	292	90	157	135	180	135	158	180	180	337
	1.5	1.4	1.6	1.7	1.4	1.5	1.5	1.7	1.7	1.6	1.0	1.4
Prescott	202	202	225	203	203	225	225	203	203	202	202	202
	1.7	1.4	1.7	2.0	2.7	2.3	1.9	1.6	2.0	2.1	2.2	1.5
Tucson	113	292	292	90	248	292	113	113	113	113	113	113
	2.6	1.7	1.6	1.4	1.7	1.2	1.5	1.6	2.1	2.3	3.4	2.6
Yuma	0	0	337	293	135	157	157	157	158	337	0	0
	2.6	2.1	1.5	1.5	1.7	2.2	3.0	2.5	1.9	1.4	2.4	2.7
California												
Arcata	135	315	315	315	315	315	315	315	315	336	135	135
	2.3	2.8	3.3	3.9	4.4	6.6	6.9	5.0	3.6	2.1	3.1	2.5
Bishop	0	0	0	0	0	180	158	180	180	180	0	0
	4.1	4.0	3.2	2.6	2.0	2.4	2.1	2.3	2.5	3.6	5.7	3.9
Blythe	0	0	0	225	180	225	180	180	180	203	0	337
	2.3	2.4	2.2	2.3	1.4	2.1	1.9	3.9	1.8	1.9	4.2	1.9
China Lake	203	225	225	225	225	225	225	203	225	225	225	225
	1.6	1.7	1.7	1.6	1.8	2.0	2.1	2.4	2.4	1.9	1.6	1.7
El Centro	248	248	248	248	248	270	270	270	270	270	270	248
	1.6	2.0	2.9	3.6	3.8	3.7	2.1	2.3	2.6	4.1	1.6	1.4
Fort Bragg	113	315	315	315	337	315	338	337	337	135	113	135
	2.3	2.3	2.3	4.3	4.1	4.5	3.9	3.7	4.5	3.6	1.7	1.6
Fresno	157	315	315	315	315	293	293	293	315	158	135	315
	2.7	2.9	3.1	2.6	2.2	4.6	4.2	3.1	2.4	1.9	2.0	2.0
Marysville	157	135	315	135	135	135	135	157	315	135	315	338
	3.3	4.3	3.0	3.2	6.4	4.6	3.6	3.2	4.9	4.6	9.3	1.9
Merced	135	135	337	337	337	337	337	337	337	337	315	135
	3.8	3.5	3.3	4.2	3.3	3.6	4.0	4.0	3.6	3.6	3.5	4.1
Palmdale	247	225	247	225	225	225	225	225	225	225	248	225
	1.8	2.1	1.6	2.0	2.4	4.2	5.7	5.0	4.3	2.4	1.9	1.7
Riverside	23	338	293	293	293	293	315	293	293	293	67	69
	1.2	1.2	1.6	2.0	2.5	2.8	3.6	3.7	3.1	2.3	1.1	1.4
Salinas	135	135	135	293	292	292	292	315	315	315	135	135
	6.8	4.5	3.5	2.8	3.0	3.3	3.4	3.2	4.2	3.0	4.2	5.3
San Diego	180	180	180	202	248	203	293	248	293	203	180	337
	1.5	1.5	1.4	1.3	1.1	1.2	1.3	1.1	1.5	1.4	1.4	1.3
San Miguel I	157	157	337	337	315	315	293	315	315	293	294	337
	2.8	4.0	2.6	4.0	3.2	4.1	5.4	2.6	2.6	4.6	3.5	3.6
Santa Rosa	315	0	202	315	202	159	180	158	180	180	180	0
	3.1	1.9	1.6	1.6	1.4	1.4	2.7	2.1	1.6	1.7	1.5	4.8
Stockton	235	316	315	315	270	270	270	270	292	315	135	135
	4.0	2.2	2.5	2.0	1.7	2.4	3.1	2.9	2.5	3.7	2.1	2.3

Table 2 (continued)
PREVAILING WIND-EROSION DIRECTION[a] (FIRST LINE) AND
PREPONDERANCE OF PREVAILING WIND-EROSION FORCES IN
THE PREVAILING WIND-EROSION DIRECTION (SECOND LINE)

Location	Jan	Feb	Mar	Apr	May	Jun	Jul	Aug	Sep	Oct	Nov	Dec
Thermal	337	157	337	337	337	337	337	157	337	337	337	315
	2.4	2.6	2.3	4.3	3.0	4.4	3.7	4.3	4.1	2.6	2.8	3.4
Victorville	193	270	251	270	270	225	180	180	180	248	248	248
	1.1	1.2	1.3	1.4	1.3	1.1	2.0	1.8	1.4	1.2	1.3	1.4
Colorado												
Denver	315	315	293	338	338	159	338	0	158	0	270	0
	1.2	1.1	1.0	1.2	1.2	1.2	1.8	1.3	1.6	1.1	1.3	1.3
La Junta	248	292	247	0	45	225	67	203	46	248	0	270
	1.5	1.1	1.3	1.1	1.1	1.5	1.0	1.4	1.3	1.3	1.3	1.3
Pueblo	293	337	270	202	225	203	0	337	23	293	292	293
	1.6	1.2	1.6	1.3	1.4	1.3	1.2	1.1	1.2	1.1	1.4	1.5
Connecticut												
Bridgeport	293	315	293	293	90	225	270	247	45	248	337	293
	1.2	1.3	1.2	1.5	1.2	1.6	1.3	1.7	1.5	1.3	1.6	1.4
Hartford	315	315	315	338	0	180	180	180	180	0	315	315
	1.8	1.8	1.5	2.0	1.5	1.6	1.4	1.5	1.8	1.4	1.3	1.6
Stratford	315	293	293	225	225	202	—	204	225	247	292	270
	1.3	2.3	1.4	1.4	1.6	1.6	—	1.3	2.5	2.5	1.5	1.6
Windsor Lock	315	315	337	315	180	180	180	180	0	315	337	315
	1.8	4.6	1.3	1.9	2.1	2.0	1.6	1.7	1.7	1.5	1.9	2.0
Delaware												
Wilmington	315	293	315	315	338	315	158	180	180	68	315	293
	1.7	2.0	1.8	1.3	1.1	1.2	1.4	1.3	1.4	1.1	1.7	1.4
Florida												
Avon Park	338	22	180	45	67	270	337	68	68	45	0	0
	1.5	1.6	1.2	1.6	2.0	1.2	1.2	1.7	1.8	2.1	1.5	2.4
Cape	337	315	157	135	112	135	135	135	90	45	337	335
Canaveral	1.5	1.4	1.4	1.4	1.3	1.3	1.6	1.3	1.1	1.4	1.3	1.4
Daytona	202	225	225	68	68	68	90	67	67	45	0	0
Beach	1.2	1.2	1.1	1.3	1.6	1.7	1.6	1.7	1.7	2.0	1.4	1.3
Fort Myers	23	225	45	67	68	90	281	68	67	45	45	22
	1.4	1.3	1.3	1.4	2.0	1.2	1.2	1.6	1.7	1.9	1.8	2.1
Homestead	112	112	90	90	90	90	112	90	90	65	67	112
	1.2	1.4	1.3	2.0	2.4	2.5	2.5	2.5	1.7	2.2	1.6	1.3
Jacksonville	225	247	270	270	270	68	90	45	45	45	225	225
	1.3	1.5	1.5	1.6	1.8	1.7	1.5	1.5	1.9	1.9	1.2	1.3
Key West	0	135	135	113	112	113	135	112	90	45	45	23
	1.1	1.3	1.6	1.3	2.1	1.8	1.7	2.2	1.2	1.9	1.5	1.5
Orlando	315	315	315	114	270	270	158	270	69	67	0	338
	1.2	1.1	1.1	1.1	1.4	1.2	1.1	1.3	1.3	1.3	1.4	1.2
Pensacola	180	202	180	135	158	180	180	203	22	157	0	337
	1.6	1.5	1.8	1.6	1.6	1.8	1.5	1.3	1.1	1.6	2.0	1.5
Perry Field	180	203	225	202	225	225	225	225	225	315	338	180
	1.8	3.8	1.6	3.0	3.2	4.0	2.5	2.2	2.2	1.3	2.3	1.3
Sarasota	0	0	22	202	249	225	337	203	45	22	22	0
	1.6	1.6	1.6	1.3	1.4	1.3	1.3	1.3	2.3	1.7	1.6	1.6
Tallahassee	338	180	180	180	180	180	180	247	45	0	338	338
	1.8	1.5	1.7	1.7	1.4	1.2	1.5	1.2	1.5	1.2	1.9	2.1
Tampa	0	338	202	248	270	270	292	248	45	45	23	22
	1.3	1.2	1.0	1.1	1.3	1.3	1.2	1.3	1.4	2.0	1.7	1.6
Vero Beach	113	112	135	68	67	68	270	47	67	67	67	292
	1.5	1.2	1.3	1.3	1.6	1.6	1.6	1.1	1.8	2.4	1.1	1.2

Table 2 (continued)
PREVAILING WIND-EROSION DIRECTION[a] (FIRST LINE) AND
PREPONDERANCE OF PREVAILING WIND-EROSION FORCES IN
THE PREVAILING WIND-EROSION DIRECTION (SECOND LINE)

Location	Jan	Feb	Mar	Apr	May	Jun	Jul	Aug	Sep	Oct	Nov	Dec
West Palm	113	113	135	113	90	112	113	112	68	67	68	90
Beach	1.3	1.5	1.6	1.6	1.6	1.5	2.3	1.8	1.8	2.1	1.4	1.3
Georgia												
Albany	178	180	158	135	222	247	225	225	45	45	293	296
	1.3	1.2	1.3	1.2	1.1	1.4	1.4	1.2	1.6	1.5	1.3	1.1
Athens	270	90	112	248	248	248	225	247	67	67	90	270
	1.4	1.4	1.5	1.4	1.6	1.3	1.5	1.5	2.0	1.8	1.5	1.4
Bainbridge	225	160	158	225	67	157	225	225	67	45	315	270
	1.1	1.0	1.4	1.2	1.1	1.1	1.5	1.3	1.9	1.3	1.5	1.2
Marietta	315	293	315	293	295	292	270	293	90	293	315	293
	1.5	1.5	1.5	1.5	1.3	1.4	1.2	1.4	1.5	1.5	1.9	1.7
Savannah	292	270	270	112	270	270	203	113	68	45	270	270
	1.5	1.6	1.3	1.2	1.2	1.2	1.2	1.1	1.2	1.9	1.3	1.4
Valdosta	270	225	180	203	247	68	225	45	67	67	270	247
	1.2	1.0	1.1	1.2	1.2	1.4	1.4	1.7	2.3	1.6	1.3	1.3
Waycross	293	293	90	90	113	90	0	67	67	67	0	23
	1.3	1.7	1.2	1.2	1.7	1.7	1.2	3.0	2.6	2.1	2.1	1.2
Idaho												
Boise	135	315	315	315	315	315	315	293	315	113	135	135
	3.4	4.9	2.9	3.1	2.6	2.7	2.2	1.7	2.7	2.3	3.5	3.0
Mountain	113	135	293	113	113	113	113	135	113	135	113	113
Home	2.2	2.6	2.4	2.5	2.3	2.0	2.1	2.1	2.3	3.0	2.4	2.3
Pocatello	180	203	247	247	247	225	225	225	247	247	225	203
	2.0	1.5	2.2	2.6	1.8	1.6	1.3	1.3	2.1	1.5	1.3	1.6
Illinois												
Belleville	315	315	293	315	315	90	248	338	2	338	315	315
	1.9	1.9	1.6	1.3	1.4	1.0	1.1	1.0	1.1	1.3	1.4	1.6
Glenview	225	246	23	203	23	203	23	23	203	203	203	225
	1.3	1.1	1.7	1.7	2.3	2.0	2.0	2.5	2.0	1.8	1.4	1.3
Lawrenceville	315	338	225	315	203	203	225	35	180	22	157	292
	1.3	1.2	1.6	1.1	1.8	1.7	1.8	1.6	1.8	1.3	1.2	1.2
Park Ridge	90	292	270	248	225	225	247	225	225	247	270	248
	1.2	1.3	1.4	1.2	1.4	1.2	1.3	1.5	1.2	1.4	1.2	1.2
Peoria	337	293	270	293	180	202	180	180	180	338	315	315
	1.3	1.4	1.2	1.3	1.2	1.3	1.3	1.5	1.7	1.5	1.3	1.2
Rantoul	203	315	248	203	202	224	225	202	202	180	180	158
	1.1	1.1	1.1	1.2	1.4	1.3	1.5	1.6	1.5	1.4	1.2	1.2
Rockford	247	270	248	248	225	247	248	203	203	203	244	292
	1.2	1.4	1.6	1.3	1.2	1.9	1.3	1.6	1.4	1.1	1.1	1.2
Indiana												
Bunker Hill	248	270	225	247	225	224	225	203	203	238	247	226
	1.3	1.7	1.5	1.3	1.5	1.4	1.6	1.6	1.4	1.3	1.5	1.4
Columbus	0	202	248	180	202	203	225	23	180	180	315	180
	1.3	1.1	1.0	1.1	1.5	1.4	1.2	1.4	1.6	1.6	1.1	1.1
Madison	249	270	292	248	158	202	248	202	338	338	247	270
	1.5	1.4	1.4	1.1	1.1	1.2	1.4	1.3	1.5	1.2	1.2	1.6
South Bend	225	270	90	315	338	338	338	0	180	180	225	225
	1.2	1.2	1.3	1.1	1.1	1.3	1.2	1.3	1.3	1.2	1.4	1.2
Iowa												
Burlington	293	293	270	293	248	202	225	0	315	202	315	292
	1.8	1.3	1.2	1.4	1.3	1.3	1.2	1.3	1.2	1.6	1.3	1.1
Des Moines	315	315	315	315	338	203	315	338	315	337	315	315
	1.5	1.1	1.4	1.6	1.3	1.1	1.3	1.2	1.1	1.5	1.7	1.3

Table 2 (continued)
PREVAILING WIND-EROSION DIRECTION[a] (FIRST LINE) AND PREPONDERANCE OF PREVAILING WIND-EROSION FORCES IN THE PREVAILING WIND-EROSION DIRECTION (SECOND LINE)

Location	Jan	Feb	Mar	Apr	May	Jun	Jul	Aug	Sep	Oct	Nov	Dec
Kansas												
Dodge City	0	0	0	0	180	180	180	180	180	180	0	0
	2.1	2.0	2.4	1.7	2.2	2.4	2.1	2.5	2.8	2.5	2.2	2.3
Goodland	338	338	338	338	180	180	180	180	180	158	337	338
	2.0	2.5	2.5	2.1	1.9	1.9	2.0	2.2	2.7	2.4	2.3	2.4
Olathe	180	180	180	180	180	202	202	202	180	180	180	180
	1.7	1.6	1.6	1.3	1.6	1.8	2.2	1.9	2.0	2.0	1.6	1.5
Salina	0	0	0	180	180	180	180	180	180	180	338	0
	2.0	1.9	1.8	1.6	1.8	2.0	2.1	2.0	2.7	2.1	1.8	1.7
Topeka	180	338	180	338	225	0	0	338	180	180	338	180
	1.5	1.5	1.6	1.4	1.2	1.7	1.5	1.2	2.0	2.2	1.5	1.6
Wichita	0	180	180	180	180	180	180	180	180	180	180	180
	2.5	2.1	1.6	2.1	2.1	2.3	2.1	1.9	2.3	2.9	2.1	1.8
Maine												
Bangor	337	337	337	338	338	0	338	337	338	338	338	315
	1.9	1.7	1.7	1.5	1.6	1.6	1.5	1.4	1.6	1.6	1.7	1.9
Brunswick	0	338	338	0	180	202	180	202	202	0	0	0
	1.9	1.6	1.5	1.6	1.9	1.8	1.5	1.8	2.0	1.5	1.4	1.5
Caribou	293	292	293	293	315	315	293	293	292	293	292	292
	1.8	1.6	1.9	1.7	1.5	1.2	1.3	1.4	1.3	1.6	1.6	1.6
Maryland												
Aberdeen	315	315	315	315	0	202	202	22	22	0	315	315
	1.3	1.9	1.6	1.3	1.3	1.3	1.6	1.7	1.4	1.4	1.5	1.5
Annapolis	315	315	315	338	180	180	180	180	180	180	315	315
	1.8	1.9	1.7	1.5	1.5	1.8	2.0	2.0	2.0	1.2	1.4	1.7
Frederick	293	315	315	315	315	315	293	315	315	315	293	293
	2.6	2.9	4.0	2.1	1.9	1.9	2.3	1.3	1.8	1.8	2.6	2.8
Massachusetts												
Bedford	293	293	292	292	270	247	225	225	23	248	293	292
	1.4	1.5	1.3	1.2	1.1	1.4	1.4	1.3	1.2	1.3	1.1	1.5
Chicopee Falls	337	315	338	338	0	180	180	180	0	0	0	315
	1.5	2.0	1.6	1.6	1.6	1.9	2.1	2.1	1.9	1.7	1.5	1.6
Nantucket	315	293	292	23	225	225	225	204	23	45	315	292
	1.4	1.4	1.3	1.3	1.6	1.9	1.9	1.6	1.5	1.3	1.2	1.3
Worcester	248	248	67	270	248	225	248	45	45	67	248	248
	1.9	2.1	1.5	1.7	1.7	1.8	1.8	2.2	3.2	2.2	1.4	1.6
Michigan												
Battle Creek	248	248	248	270	247	248	248	270	270	225	225	225
	1.4	1.4	1.6	1.4	1.5	1.5	1.5	1.3	1.3	1.2	1.3	1.5
Cadillac	248	248	292	292	225	225	247	225	246	203	203	247
	1.4	1.3	1.2	1.5	1.4	1.2	1.3	1.4	1.2	1.4	1.2	1.5
Flint	225	270	248	248	247	248	248	225	225	225	225	225
	1.4	1.4	1.6	1.4	1.3	1.6	1.4	1.8	1.2	1.2	1.6	1.5
Marquette	0	338	0	0	0	180	202	0	180	180	180	180
	1.8	1.8	1.9	1.7	2.1	1.8	1.9	2.9	1.7	2.0	1.8	1.8
Mt. Clemens	225	225	225	203	180	201	202	180	180	202	203	225
	1.5	1.2	1.2	1.3	1.3	1.3	1.4	1.5	1.6	1.5	1.4	1.4
Muskegon	248	270	248	225	205	225	225	203	203	203	225	270
	1.7	1.6	1.4	1.2	1.5	1.7	1.4	2.3	1.4	1.4	1.5	1.1
Oscoda	338	315	270	239	227	270	202	225	248	224	226	315
	1.2	1.2	1.2	1.3	1.1	1.1	1.1	1.3	1.2	1.2	1.1	1.0
Pellston	270	270	270	270	270	248	248	248	248	248	292	270
	1.4	1.7	1.6	1.5	1.5	1.8	2.0	1.7	1.5	1.3	1.4	1.4

Table 2 (continued)
PREVAILING WIND-EROSION DIRECTION[a] (FIRST LINE) AND
PREPONDERANCE OF PREVAILING WIND-EROSION FORCES IN
THE PREVAILING WIND-EROSION DIRECTION (SECOND LINE)

Location	Jan	Feb	Mar	Apr	May	Jun	Jul	Aug	Sep	Oct	Nov	Dec
Sault Saint	292	293	293	293	293	293	293	293	293	293	293	292
	1.8	2.3	2.5	2.9	2.5	2.6	3.1	2.2	2.3	2.3	1.9	2.1
Traverse City	203	202	202	202	203	225	203	203	202	202	180	225
	1.3	1.3	1.4	1.4	1.6	1.7	1.6	1.6	1.5	1.7	1.7	1.3
Ypsilanti	248	270	270	270	270	270	270	270	270	248	248	248
	1.5	1.6	1.9	1.8	1.5	1.6	1.6	1.3	1.5	1.3	1.6	1.5
Minnesota												
Duluth	292	270	293	90	90	248	270	68	270	248	293	293
	1.9	1.6	1.3	1.7	1.7	2.0	1.7	1.5	1.8	1.3	1.6	1.7
Int. Falls	292	270	293	292	292	293	292	315	315	293	293	292
	1.4	1.5	1.3	1.4	1.3	1.4	1.5	1.3	1.3	2.3	1.5	1.9
Minneapolis	315	293	292	293	292	293	180	180	315	315	315	293
	1.5	1.7	1.4	1.5	1.3	1.3	1.0	1.2	1.4	1.6	1.7	1.4
Rochester	337	315	315	315	157	158	157	180	180	157	315	337
	2.0	1.9	1.7	1.8	1.6	1.8	1.7	2.1	2.0	2.1	1.6	1.6
Montana												
Billings	203	202	315	0	315	0	0	338	338	338	316	291
	1.4	1.2	1.3	1.3	1.3	1.1	1.4	1.4	1.5	1.2	1.0	1.2
Cut bank	270	270	270	270	270	247	270	270	270	270	248	270
	4.5	3.3	2.2	2.3	1.9	1.3	2.5	2.5	2.1	1.9	2.6	4.4
Glasgow	293	294	293	293	293	112	292	113	293	293	293	293
	2.9	3.0	2.8	2.1	2.2	2.2	2.3	2.3	2.6	3.9	3.6	3.6
Great Falls	225	225	225	247	248	248	247	248	247	225	225	225
	3.2	3.3	1.9	1.6	1.8	1.6	1.6	1.5	1.9	2.1	3.5	3.6
Helena	180	270	292	293	292	292	293	293	293	270	248	293
	1.4	1.8	1.6	1.5	1.6	1.8	1.4	1.4	1.7	1.2	1.3	1.2
Lewistown	270	270	270	292	293	293	315	292	293	293	225	226
	1.7	1.4	2.1	1.9	2.7	2.0	1.5	2.1	1.6	1.5	1.5	1.6
Miles City	315	293	293	315	293	292	315	315	315	293	293	293
	1.8	1.7	2.1	2.6	2.0	1.8	1.6	1.6	1.9	2.3	3.1	2.5
Missoula	112	292	292	248	270	270	293	315	315	338	293	113
	1.5	1.4	1.5	1.4	1.2	1.3	1.5	1.2	1.2	1.5	1.6	1.6
Nebraska												
Grand Island	338	338	338	0	180	180	180	180	180	158	338	338
	2.2	2.1	1.6	1.6	1.7	2.0	1.6	1.9	2.2	2.4	1.9	2.1
Lincoln	338	337	338	158	337	158	158	158	158	338	337	338
	2.4	1.5	1.5	1.6	1.6	1.6	1.8	1.5	1.8	1.7	2.1	2.1
North Platte	338	338	338	338	338	158	158	158	180	338	338	338
	2.3	2.6	2.1	2.6	2.2	2.4	2.0	2.6	2.4	3.1	3.0	3.4
Omaha	337	337	337	315	180	158	169	158	158	338	337	338
	2.9	1.7	1.7	1.9	1.7	1.6	1.7	1.3	2.0	1.9	2.1	2.1
Scotts Bluff	292	293	315	315	337	294	135	315	0	315	337	292
	2.1	2.0	1.9	1.6	1.7	1.6	1.4	1.8	1.3	2.3	2.9	2.7
Nevada												
Fallon	225	203	247	270	270	292	292	248	315	247	159	180
	1.6	1.5	1.2	1.4	1.2	1.8	1.5	1.5	1.1	1.5	1.2	1.5
Las Vegas	180	0	203	225	203	225	225	203	225	203	22	0
	1.8	1.4	1.7	1.9	2.2	2.6	2.0	2.0	2.6	1.9	1.8	1.8
Mercury	180	0	202	180	202	180	181	180	180	180	180	180
	2.8	2.1	2.2	2.2	2.3	2.9	1.3	3.2	2.8	2.8	3.2	3.7
Reno	180	180	158	157	293	293	293	292	315	158	180	180
	2.2	2.1	1.3	1.6	1.6	2.4	2.4	2.7	1.2	1.5	2.1	2.2
Tonopah	157	315	315	315	315	157	158	180	158	315	315	337
	3.0	3.0	3.2	2.7	2.3	2.3	2.1	2.6	2.6	2.5	3.1	2.6

Table 2 (continued)
PREVAILING WIND-EROSION DIRECTION[a] (FIRST LINE) AND PREPONDERANCE OF PREVAILING WIND-EROSION FORCES IN THE PREVAILING WIND-EROSION DIRECTION (SECOND LINE)

Location	Jan	Feb	Mar	Apr	May	Jun	Jul	Aug	Sep	Oct	Nov	Dec
New Hampshire												
Concord	315	315	315	315	315	315	315	315	315	315	315	315
	2.8	2.8	2.2	2.0	2.2	1.8	2.1	1.8	1.6	1.9	2.0	3.0
Manchester	315	315	294	293	315	315	158	315	337	337	315	292
	1.7	1.6	1.4	1.6	1.3	1.2	1.3	1.5	1.4	1.5	1.5	1.8
Portmouth	315	293	293	293	293	292	315	337	158	315	315	315
	1.4	1.9	1.6	1.4	1.4	1.2	1.4	1.2	1.1	1.4	1.4	1.4
New Jersey												
Atlantic City	292	292	292	292	293	304	180	202	315	270	293	293
	2.2	2.0	1.9	1.6	1.3	1.2	1.2	1.3	1.2	1.1	1.7	1.6
Lakehurst	270	270	270	248	248	247	225	225	225	247	248	270
	1.8	2.0	1.9	1.5	1.7	1.8	2.1	1.7	1.7	1.8	1.6	1.8
Trenton	292	293	292	292	292	249	225	225	22	67	293	292
	1.2	1.6	1.4	1.2	1.2	1.2	1.4	1.1	1.4	1.2	1.5	1.4
New Mexico												
Albuquerque	112	293	293	113	112	135	135	113	113	135	112	113
	1.8	1.6	1.5	1.4	1.3	1.5	1.4	1.9	1.4	1.3	1.4	1.5
Hobbs	248	270	247	270	247	180	180	202	203	202	225	247
	1.8	1.5	2.1	1.3	1.2	1.3	1.6	1.8	2.7	1.8	1.7	1.3
Roswell	202	190	203	202	180	180	180	158	180	180	202	180
	1.6	1.5	1.3	1.7	1.6	2.0	2.2	2.0	2.6	2.2	1.4	1.5
New York												
Geneva	315	293	157	0	337	338	202	180	180	180	180	180
	2.0	1.9	1.8	2.4	2.4	1.8	2.8	2.1	2.3	2.1	1.8	1.4
New York	292	292	292	292	292	157	158	180	158	270	270	292
	1.6	1.9	1.7	1.5	1.1	1.4	1.4	1.2	1.1	1.1	1.5	1.7
Niagara	247	247	226	225	225	225	225	225	225	225	225	247
	1.7	1.5	1.6	2.0	1.6	1.8	2.1	2.2	1.7	1.7	1.8	2.0
Plattsburgh	293	315	315	315	135	135	315	135	157	135	157	157
	1.3	1.3	1.6	1.6	2.0	1.8	1.6	2.1	1.8	2.1	1.3	1.5
Rome	292	293	293	293	293	315	293	293	293	315	293	292
	2.7	3.1	2.6	2.5	2.5	2.1	1.7	2.1	1.9	1.7	2.2	2.5
Schenectady	293	293	293	293	293	292	292	293	293	293	292	293
	5.6	2.8	2.4	1.6	2.6	2.2	2.5	1.9	1.5	2.1	2.0	2.7
West Hampton	315	293	292	248	225	225	225	225	225	67	0	292
	1.2	1.5	1.2	1.2	1.2	1.4	1.8	1.6	1.6	1.3	1.1	1.2
North Carolina												
Cherry Point	23	203	203	203	203	203	203	23	23	23	23	23
	1.7	1.7	2.0	2.6	2.9	2.5	2.8	2.2	2.5	2.3	2.0	1.7
Hatteras	0	22	202	202	203	203	203	203	23	23	0	22
	1.2	1.5	1.4	1.8	2.8	2.6	2.9	1.6	1.7	1.9	1.6	1.3
Jacksonville	337	315	225	207	203	180	203	45	26	23	1	0
	1.3	1.3	1.2	1.6	1.9	1.5	2.1	2.0	1.7	2.4	1.2	1.3
Weeksville	22	202	225	203	225	23	45	45	45	22	338	23
	1.9	2.0	1.5	2.7	1.7	1.6	3.1	2.4	3.8	2.0	3.1	1.6
Wilmington	180	180	338	202	202	180	202	202	0	0	0	0
	1.4	1.5	1.1	1.3	2.0	2.1	2.5	1.6	1.4	2.0	2.0	1.4
Winston-	225	247	225	225	225	45	225	45	45	45	45	225
Salem	1.2	1.2	1.2	1.4	2.3	2.2	2.2	2.3	2.8	2.2	1.5	1.3
North Dakota												
Bismarck	315	293	325	315	315	293	315	157	315	315	315	315
	1.7	2.0	1.3	1.7	1.8	1.7	1.5	1.4	2.1	2.0	2.4	2.4
Fargo	337	337	337	315	337	315	157	158	315	315	337	337
	2.5	2.0	1.9	1.7	1.5	1.2	1.7	1.6	1.6	1.9	2.4	2.2

Table 2 (continued)
PREVAILING WIND-EROSION DIRECTION[a] (FIRST LINE) AND PREPONDERANCE OF PREVAILING WIND-EROSION FORCES IN THE PREVAILING WIND-EROSION DIRECTION (SECOND LINE)

Location	Jan	Feb	Mar	Apr	May	Jun	Jul	Aug	Sep	Oct	Nov	Dec
Grand Forks	338	337	338	337	337	270	337	158	315	337	337	337
	3.2	2.5	2.1	2.0	1.5	1.2	1.4	1.7	1.3	1.8	2.5	2.5
Minot	315	315	315	315	315	293	315	337	293	315	315	315
	1.7	1.8	1.7	1.5	1.5	1.3	1.5	1.4	1.7	1.8	2.2	1.6
Ohio												
Cincinnati	203	225	247	225	225	225	225	225	225	225	225	225
	1.5	1.3	1.3	1.6	1.6	1.8	1.7	1.8	1.8	1.7	1.5	1.6
Cleveland	248	270	247	247	203	225	203	293	202	203	247	225
	1.3	1.2	1.3	1.3	1.3	1.2	1.4	1.1	1.2	1.4	1.3	1.2
Columbus	315	270	247	225	203	202	202	0	180	180	225	203
	1.1	1.1	1.2	1.3	1.4	1.4	1.4	1.4	1.4	1.4	1.3	1.2
Daytona	292	270	270	248	247	225	246	225	203	224	247	225
	1.1	1.3	1.4	1.3	1.3	1.3	1.4	1.2	1.2	1.2	1.2	1.1
Toledo	247	247	248	247	247	225	204	225	248	225	220	225
	1.4	1.5	1.4	1.6	1.7	1.4	1.1	1.3	1.4	1.5	1.6	2.0
Youngstown	248	270	248	248	248	247	225	202	225	225	247	247
	1.4	1.5	1.6	1.4	1.2	1.1	1.3	1.3	1.3	1.3	2.0	1.7
Oklahoma												
Oklahoma City	0	180	180	180	180	180	180	180	180	180	180	180
	2.6	2.4	1.7	2.4	2.1	2.1	2.1	2.0	2.1	3.1	2.6	2.5
Tulsa	0	180	0	0	180	180	180	180	180	0	0	180
	2.6	2.8	1.8	2.2	2.2	2.7	2.0	1.9	2.8	2.7	2.7	2.1
Oregon												
Astoria	203	202	203	202	315	315	315	315	180	203	202	203
	1.6	1.6	1.5	1.2	1.3	1.7	3.1	2.1	1.4	2.3	2.1	1.8
Klamath Falls	180	158	157	157	337	320	337	337	158	159	158	158
	1.8	1.8	1.3	1.5	1.6	1.3	2.0	1.8	1.4	1.9	2.0	1.4
Medford	157	136	157	315	315	293	315	315	315	157	152	158
	1.5	1.5	1.5	1.6	2.3	1.7	1.7	2.2	1.7	1.7	1.9	2.6
Pendleton	248	270	270	270	270	270	270	270	270	270	270	270
	1.2	2.0	3.1	4.1	4.8	4.2	3.2	3.9	3.7	3.1	2.7	1.7
Portland	180	202	202	180	180	158	315	315	180	180	180	180
	1.1	1.5	1.2	1.5	1.4	1.3	1.8	1.8	1.2	1.6	1.1	1.2
Redmond	180	158	315	315	315	315	315	315	337	180	180	180
	1.6	1.8	1.6	1.7	2.6	1.9	2.7	2.7	3.2	1.7	1.5	2.4
Salem	180	180	180	180	180	202	0	0	180	180	180	180
	6.6	5.4	3.8	2.1	1.6	1.4	1.7	1.8	2.6	4.5	5.6	5.5
Rhode Island												
Quonset Point	0	338	0	0	202	202	202	180	0	22	0	338
	1.6	1.3	1.5	1.6	2.0	2.2	2.2	2.1	2.2	2.0	1.5	1.3
South Carolina												
Beaufort	247	270	292	180	270	23	202	225	45	45	202	225
	1.2	1.7	1.6	1.3	1.4	1.5	1.9	1.3	1.6	2.3	1.2	1.3
Florence	203	225	225	225	225	203	203	203	45	23	23	225
	2.1	2.3	1.6	1.5	1.4	2.3	1.8	2.2	1.8	2.1	1.9	1.8
Greenville	45	45	225	225	45	45	45	45	45	45	45	45
	2.0	2.4	2.0	1.9	2.4	2.1	2.0	2.3	3.3	2.7	2.3	2.3
Myrtle Beach	225	225	225	203	203	202	203	202	225	45	247	247
	1.3	1.2	1.2	1.3	1.8	1.7	2.3	2.0	1.4	1.3	1.2	1.4
Sumter	225	225	225	225	225	203	225	45	45	45	225	225
	1.7	1.8	1.5	1.6	1.6	1.4	1.9	1.7	3.1	2.3	1.9	1.7

Table 2 (continued)
PREVAILING WIND-EROSION DIRECTION[a] (FIRST LINE) AND PREPONDERANCE OF PREVAILING WIND-EROSION FORCES IN THE PREVAILING WIND-EROSION DIRECTION (SECOND LINE)

Location	Jan	Feb	Mar	Apr	May	Jun	Jul	Aug	Sep	Oct	Nov	Dec
South Dakota												
Huron	337	337	337	337	157	158	158	158	158	315	315	315
	2.9	2.6	2.6	1.8	1.8	1.9	2.3	2.6	2.0	2.4	2.5	2.7
Rapid City	337	337	337	337	337	337	337	337	337	337	337	315
	2.7	2.9	3.0	2.4	2.0	2.4	2.4	2.3	2.7	2.6	3.0	2.9
Sioux Falls	337	338	315	337	315	315	158	158	337	338	315	337
	2.2	1.5	1.5	1.9	1.4	1.4	1.6	1.6	1.6	1.7	2.3	2.0
Texas												
Amarillo	203	0	225	203	203	202	180	202	202	202	202	23
	1.9	1.7	1.3	1.4	1.3	1.9	1.4	1.9	2.1	2.0	1.7	1.8
Austin	0	0	346	0	180	180	180	0	0	0	0	0
	2.3	1.8	1.7	1.8	2.1	1.9	1.7	1.9	1.9	2.1	2.4	1.8
Brownsville	157	157	157	135	135	135	135	135	157	135	337	158
	3.6	2.6	2.1	2.1	3.1	3.0	3.7	2.5	1.4	1.7	3.2	2.9
Corpus Christi	158	157	158	157	157	157	157	135	135	157	158	338
	3.1	3.0	2.4	2.3	2.3	3.4	2.7	1.8	1.3	1.7	2.5	2.3
Dalhart	225	225	225	203	225	203	202	202	203	203	203	23
	2.4	2.2	2.3	1.8	2.0	2.2	2.3	2.3	2.8	2.6	2.0	2.2
Dallas	180	158	158	180	180	180	180	180	180	180	180	178
	2.9	2.2	1.9	2.3	2.2	2.6	2.1	1.9	1.9	2.5	1.8	2.1
Galveston	337	337	337	135	157	158	180	180	41	135	0	338
	1.9	1.5	1.4	1.3	1.9	2.2	2.3	1.5	1.4	1.2	2.0	1.3
Houston	158	158	158	157	157	180	180	158	45	68	0	338
	1.9	1.6	1.6	1.8	1.7	1.9	1.8	1.6	1.2	1.4	1.6	1.6
Laredo	155	135	135	135	135	135	135	135	135	135	157	157
	1.9	1.8	1.7	2.4	2.9	3.5	3.8	3.2	1.7	2.2	1.8	2.0
Lubbock	227	247	248	225	202	180	180	180	202	203	214	248
	1.3	1.2	1.5	1.1	1.3	1.9	2.0	1.5	2.0	1.6	1.2	1.2
Midland	247	270	225	293	180	180	180	180	180	180	225	0
	1.6	1.1	1.2	1.1	1.5	1.8	1.7	1.5	1.8	1.9	1.4	1.1
Port Arthur	315	180	180	158	158	180	180	22	23	338	158	158
	1.5	1.6	2.2	2.1	2.3	1.6	1.2	1.3	1.3	1.3	1.4	1.4
San Angelo	225	225	204	203	180	180	180	180	202	202	23	225
	1.5	1.3	1.1	1.5	1.8	2.0	2.8	1.6	1.8	1.7	2.1	2.7
San Antonio	23	337	338	21	135	135	135	135	45	22	0	0
	1.6	1.3	1.2	1.1	1.7	1.6	1.9	1.7	1.2	1.5	2.1	1.6
Waco	0	0	180	180	180	180	202	202	202	22	0	0
	2.5	2.4	2.5	2.4	2.5	3.0	2.4	2.3	2.0	2.4	2.5	2.6
Wichita Falls	1	0	338	180	158	180	180	180	180	180	338	0
	1.9	1.6	1.5	1.8	1.9	2.3	1.5	1.4	1.8	2.1	1.6	1.6
Utah												
Dugway	180	0	0	0	0	180	180	180	0	180	0	180
	4.0	3.2	2.8	1.9	2.0	2.1	1.8	2.2	2.0	2.2	2.6	2.6
Ogden	180	45	180	23	67	67	68	67	68	68	67	22
	1.4	1.3	1.3	1.2	1.3	1.2	1.9	1.6	2.1	2.2	1.3	1.1
Salt Lake City	158	157	158	180	337	157	180	157	158	157	158	158
	2.6	2.5	1.7	1.6	1.7	1.7	1.6	1.9	2.4	1.5	2.4	2.4
Wendover	338	315	315	315	315	337	338	157	315	315	315	315
	1.4	1.9	1.9	2.1	1.6	1.5	1.4	1.3	1.6	1.9	1.8	1.9
Virginia												
Blackstone	221	225	203	225	225	225	221	203	203	23	203	225
	1.8	1.2	1.2	1.6	1.9	1.8	2.3	1.6	1.8	1.7	1.5	1.3
Chincoteague	315	293	292	182	225	228	203	203	45	45	315	301
	1.4	1.3	1.3	1.1	1.2	1.1	1.5	1.3	1.7	1.4	1.3	1.3

Table 2 (continued)
PREVAILING WIND-EROSION DIRECTION[a] (FIRST LINE) AND PREPONDERANCE OF PREVAILING WIND-EROSION FORCES IN THE PREVAILING WIND-EROSION DIRECTION (SECOND LINE)

Location	Jan	Feb	Mar	Apr	May	Jun	Jul	Aug	Sep	Oct	Nov	Dec
Hampton	0	0	22	0	202	203	203	22	22	0	0	0
	1.5	1.3	1.4	1.5	1.5	1.7	1.9	1.7	2.1	2.1	1.6	1.5
Oceana	0	353	225	203	225	45	225	67	45	22	338	0
	1.7	1.1	1.1	1.4	1.2	1.5	2.2	1.4	1.5	1.4	1.5	1.3
Quantico	337	337	315	180	180	180	180	180	180	349	338	337
	1.7	1.4	1.3	1.2	1.4	1.2	1.6	1.5	1.7	1.3	1.5	1.5
Washington												
Everett	158	158	158	158	158	338	0	338	338	158	158	158
	2.5	2.4	2.7	2.3	2.7	3.5	2.9	3.1	4.0	2.8	2.5	2.6
Kelso	158	158	158	158	338	338	338	338	0	158	158	158
	7.7	6.3	3.8	2.3	1.8	1.6	1.9	1.8	3.3	4.5	9.2	9.2
Moses Lake	0	180	225	227	270	225	270	0	0	180	0	180
	2.5	1.9	1.3	1.1	1.1	1.2	1.1	1.2	1.3	1.6	1.6	1.9
Olympia	202	203	203	225	247	225	225	225	225	202	203	203
	3.1	2.9	2.3	2.3	2.2	2.3	2.3	2.3	2.3	2.6	2.8	3.1
Spokane	203	203	225	225	225	203	203	203	225	202	203	203
	4.1	2.8	2.5	3.1	3.4	2.1	1.9	1.6	2.6	3.4	3.9	3.5
Tacoma	203	203	203	225	225	225	225	225	203	203	202	202
	2.4	2.6	2.1	2.0	1.9	1.8	1.4	1.4	2.2	2.3	2.2	2.4
Walla Walla	202	180	202	225	225	225	225	225	225	180	180	180
	2.6	2.4	1.8	2.8	2.7	2.9	2.7	2.4	2.7	2.4	2.3	5.5
Whidbey	135	135	135	135	270	270	225	270	135	135	135	135
Island	2.7	2.5	1.9	1.7	1.4	1.7	1.6	1.2	2.2	2.7	2.8	2.8
Yakima	203	223	221	293	293	315	337	315	293	180	180	203
	2.3	1.0	1.1	1.2	1.7	1.8	1.7	2.0	1.5	1.1	1.1	1.4
Wisconsin												
Green Bay	292	228	225	247	225	225	225	225	225	225	270	227
	1.2	1.3	1.4	1.2	1.4	1.5	1.3	1.9	1.7	1.3	1.3	1.2
La Crosse	315	315	315	315	329	157	336	225	315	315	315	338
	1.6	1.9	1.8	1.8	1.1	1.2	1.4	1.2	1.5	1.5	1.9	1.0
Madison	315	270	250	270	248	247	248	225	247	247	270	248
	1.2	1.3	1.4	1.5	1.3	1.3	1.3	1.3	1.3	1.1	1.4	1.2
Milwaukee	225	225	23	225	225	203	202	225	203	180	225	247
	1.2	1.2	1.3	1.1	1.2	1.2	1.7	1.3	1.2	1.2	1.4	1.1
Wyoming												
Cheyenne	270	270	270	315	292	293	337	315	292	293	292	270
	2.4	1.7	1.2	1.5	1.4	1.3	1.5	1.2	1.4	1.7	1.9	2.3
Sheridan	315	315	315	315	315	315	315	315	315	246	315	292
	3.4	2.5	3.0	3.1	2.7	3.1	2.4	2.5	3.0	3.2	3.4	2.3

[a] Clockwise from the north through 360°.

Adapted from Skidmore and Woodruff.[17]

Procedures to determine potential average annual soil loss, $E = f(I, K, C, L, V)$ with E_1, E_2, E_3, and E_4 as intermediate steps in the solution, are as follows:

1. $E_1 = I = 166$ t/ha/year.
2. Determine $E_2 = IK = 166 (0.77) = 128$ t/ha/year.
3. Determine $E_3 = IKC = 128 (0.4) = 51$ t/ha/year.
4. Determine $E_4 = IKC f(L)$ for $L = 547$ m. Use Table 5. Interpolate between 112 and

Table 3
WIND-EROSION DIRECTION FACTOR

Angle of deviation (degrees) of prevailing wind-erosion direction from perpendicular to the field border

Preponderance	0	5	10	15	20	25	30	35	40	45	50
1.0	1.90	1.90	1.90	1.90	1.90	1.90	1.90	1.90	1.90	1.90	1.90
1.2	1.55	1.57	1.60	1.65	1.70	1.74	1.77	1.81	1.84	1.88	1.92
1.4	1.40	1.43	1.46	1.51	1.55	1.60	1.65	1.71	1.78	1.86	1.95
1.6	1.30	1.34	1.38	1.42	1.46	1.51	1.55	1.64	1.73	1.85	1.97
1.8	1.23	2.26	3.30	2.35	1.40	1.44	1.48	1.59	1.70	1.85	2.00
2.0	1.19	1.24	1.30	1.32	1.35	1.39	1.44	1.55	1.67	1.85	2.04
2.2	1.17	1.22	1.27	1.30	1.33	1.37	1.41	1.53	1.66	1.87	2.08
2.4	1.15	1.20	1.25	1.28	1.31	1.35	1.40	1.52	1.65	1.88	2.12
2.6	1.13	1.18	1.23	1.27	1.31	1.35	1.40	1.53	1.67	1.91	2.16
2.8	1.11	1.16	1.22	1.26	1.30	1.36	1.41	1.56	1.71	1.96	2.22
3.0	1.09	1.14	1.20	1.25	1.30	1.36	1.43	1.59	1.75	2.02	2.29
3.2	1.08	1.14	1.19	1.25	1.31	1.38	1.45	1.64	1.82	2.09	2.37
3.4	1.07	1.13	1.19	1.25	1.31	1.40	1.48	1.68	1.89	2.18	2.47
3.6	1.06	1.12	1.18	1.25	1.32	1.42	1.52	1.76	2.00	2.31	2.62
3.8	1.05	1.11	1.18	1.25	1.33	1.45	1.57	1.84	2.11	2.45	2.78
4.0	1.04	1.10	1.17	1.26	1.35	1.48	1.61	1.91	2.22	2.59	2.90

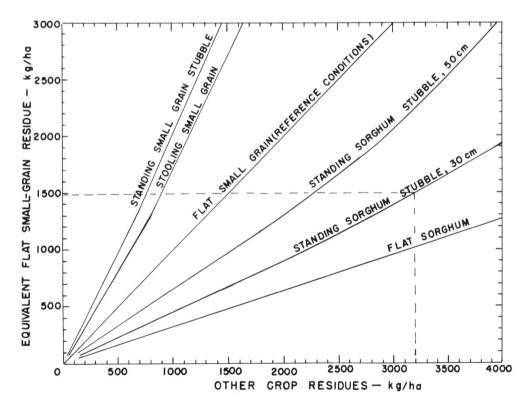

FIGURE 3.   Converting dry weight of selected crop residues to quantity of equivalent flat, small-grain residue. (From Woodruff, N. P. and Siddoway, F. H., *Soil Sci. Soc. Am. Proc.,* 29, 602, 1965. With permission.)

Table 4

COEFFICIENTS IN PREDICTION EQUATION,
$(SG)_e = aX^b$, FOR CONVERTING RANGE
GRASSES TO EQUIVALENT QUANTITY OF
FLAT, SMALL-GRAIN RESIDUE[a]

Grass species	Grazing management	Height (cm)	Prediction equation coefficients	
			a	b
Blue grama	Ungrazed	33.0	0.60	1.39
Buffalograss	Ungrazed	10.2	1.40	1.44
Big bluestem	Properly grazed	15.2	0.22	1.34
Blue grama	Properly grazed	5.1	1.60	1.08
Buffalograss	Properly grazed	5.1	3.08	1.18
Little bluestem	Properly grazed	10.2	0.19	1.37
Switchgrass	Properly grazed	15.2	0.47	1.40
Western wheatgrass	Properly grazed	10.2	1.54	1.17
Big bluestem	Overgrazed	2.5	4.12	0.92
Blue grama	Overgrazed	2.5	3.06	1.14
Buffalograss	Overgrazed	2.5	2.45	1.40
Little bluestem	Overgrazed	2.5	0.52	1.26
Switchgrass	Overgrazed	2.5	1.80	1.12
Western wheatgrass	Overgrazed	2.5	3.93	1.07

[a]   The term X is the dry weight (kg/ha) of grass species to be converted.

From Lyles, L. and Allison, B. E., *J. Range Manage.*, 32(2), 143, 1980.
With permission.

134 for E_2 and between 305 and 610 m for L to obtain about 1.6 divisions for curve deviation. Find $E_3 = 51$ t/ha/year at the bottom of Table 5. Then move right 1.6 scale divisions to obtain $E_4 \simeq 43$ t/ha/year.

5. Determine E. Use Figure 3 to convert 3200 kg/ha of standing sorghum stubble (30 cm) to about 1500 kg/ha of equivalent flat small-grain, $(SG)_e$. Use Equation 7 to compute $V = 5400$ kg/ha. Then, in Figure 4, find $E_4 = 43$ on the abscissa and move vertically upward to intersect $V = 5400$; then, move horizontally to left to ordinate $E \simeq 2.3$ t/ha/year.

This procedure can be reversed to determine factor conditions to hold E to some tolerable level. Other procedures available to solve the wind-erosion equation include graphical,[14,17] slide rule, and computer.[21,22] Further details may be found in Reference 31.

Table 5
DEVIATIONS OF CURVE OF E_2 VS. L FROM REFERENCE FOR USE IN DETERMINING E_4[21]

Soil loss E_2, tons/acre/year and tons/ha/year

Median unsheltered distance across field, L (m and ft)

Metric	English	3,048	2,438	1,829	1,219	914	610	305	244	183	122	91	61	30	24	18	12	9	6	3
		10,000	8,000	6,000	4,000	3,000	2,000	1,000	800	600	400	300	200	100	80	60	40	30	20	10
672	300	0.00	0.00	0.00	0.00	0.00	0.00	0.00	0.00	0.00	0.00	0.25	1.00	2.75	3.50	4.90	6.90	8.50	11.00	15.50
650	290	0.00	0.00	0.00	0.00	0.00	0.00	0.00	0.00	0.00	0.00	0.75	1.25	3.00	3.75	5.10	7.00	8.75	11.50	16.00
628	280	0.00	0.00	0.00	0.00	0.00	0.00	0.00	0.00	0.00	0.00	0.75	1.50	3.50	4.25	5.80	7.90	9.50	12.00	16.50
605	270	0.00	0.00	0.00	0.00	0.00	0.00	0.00	0.00	0.00	0.00	0.80	1.80	3.90	4.70	6.00	8.10	10.00	12.50	17.00
583	260	0.00	0.00	0.00	0.00	0.00	0.00	0.00	0.00	0.00	0.00	1.00	1.80	3.70	4.50	6.50	8.20	10.10	12.80	17.50
560	250	0.00	0.00	0.00	0.00	0.00	0.00	0.00	0.00	0.00	0.00	1.00	2.00	4.00	5.00	7.00	8.75	10.60	13.00	17.60
538	240	0.00	0.00	0.00	0.00	0.00	0.00	0.00	0.00	0.20	0.50	1.10	2.10	4.20	5.20	7.50	9.00	11.00	14.00	18.20
516	230	0.00	0.00	0.00	0.00	0.00	0.00	0.00	0.00	0.25	0.75	1.50	2.50	4.80	5.90	7.90	9.75	11.50	14.30	18.60
493	220	0.00	0.00	0.00	0.00	0.00	0.00	0.00	0.00	0.50	0.95	1.60	2.80	5.00	6.10	8.20	10.00	11.80	14.30	18.80
471	210	0.00	0.00	0.00	0.00	0.00	0.00	0.00	0.00	0.75	1.00	2.00	3.30	5.90	6.90	8.75	10.40	12.20	14.50	18.90
448	200	0.00	0.00	0.00	0.00	0.00	0.00	0.00	0.25	0.80	1.50	2.10	3.40	6.00	7.00	8.80	10.90	12.50	15.00	19.40
426	190	0.00	0.00	0.00	0.00	0.00	0.00	0.50	0.80	1.00	2.00	2.80	4.00	6.10	7.00	9.20	11.00	12.60	15.10	19.75
403	180	0.00	0.00	0.00	0.00	0.00	0.00	0.20	0.50	1.00	1.90	2.50	3.90	6.50	7.50	9.80	11.40	12.70	15.80	20.00
381	170	0.00	0.00	0.00	0.00	0.00	0.00	0.25	0.75	1.10	2.00	3.00	4.25	7.00	8.00	10.10	11.80	13.00	16.00	20.00
359	160	0.00	0.00	0.00	0.00	0.00	0.00	0.50	0.90	1.40	2.40	3.10	4.80	7.40	8.50	10.50	12.30	14.00	16.50	20.60
336	150	0.00	0.00	0.00	0.00	0.00	0.00	0.50	0.90	1.50	2.50	3.50	5.00	7.75	8.90	10.60	12.60	14.50	17.00	20.60
314	140	0.00	0.00	0.00	0.00	0.00	0.00	0.75	1.00	1.80	2.90	3.60	5.20	8.00	9.00	11.00	12.90	14.50	17.00	21.00
291	130	0.00	0.00	0.00	0.00	0.00	0.00	0.80	1.00	1.95	3.00	3.90	5.40	8.00	9.20	11.50	13.00	14.90	17.30	21.00
269	120	0.00	0.00	0.00	0.00	0.00	0.25	1.10	1.80	2.60	3.80	4.75	6.20	9.00	10.00	11.80	13.80	15.50	17.75	21.10
247	110	0.00	0.00	0.00	0.00	0.00	0.30	1.50	1.95	2.90	4.00	5.00	6.75	9.50	10.60	12.30	14.50	16.20	18.60	21.50
224	100	0.00	0.00	0.00	0.00	0.00	0.30	1.50	2.00	3.00	4.25	5.40	7.00	10.00	11.00	13.00	15.00	16.50	19.20	22.60
202	90	0.00	0.00	0.00	0.00	0.25	0.50	1.80	2.25	3.30	4.80	5.90	7.60	10.60	11.50	13.45	15.75	17.50	19.80	23.00
179	80	0.00	0.00	0.00	0.20	0.20	0.75	2.00	2.50	3.60	4.80	6.00	7.60	10.70	11.90	13.60	16.00	17.50	20.00	24.00
157	70	0.00	0.00	0.00	0.50	0.80	1.00	2.75	3.20	4.00	5.25	6.20	8.00	10.80	12.00	13.75	16.00	18.00	20.50	26.10
134	60	0.00	0.00	0.00	0.25	0.60	1.20	2.90	3.30	4.40	5.70	6.80	8.45	11.50	12.70	14.75	17.00	19.00	21.70	26.50
112	50	0.00	0.00	0.00	0.50	0.90	1.50	3.00	3.75	4.80	6.00	7.10	9.00	12.40	13.55	15.60	18.00	20.10	23.00	28.00
90	40	0.00	0.00	0.00	0.60	1.00	1.90	3.50	4.20	5.20	6.80	7.90	9.80	13.10	14.50	16.50	19.20	21.00	-1.00	-1.00

Table 5 (continued)
DEVIATIONS OF CURVE OF E_2 VS. L FROM REFERENCE FOR USE IN DETERMINING E_4[21]

Soil loss E_2, tons/acre/year and tons/ha/year

Median unsheltered distance across field, L (m and ft)

Metric	English	3,048 / 10,000	2,438 / 8,000	1,829 / 6,000	1,219 / 4,000	914 / 3,000	610 / 2,000	305 / 1,000	244 / 800	183 / 600	122 / 400	91 / 300	61 / 200	30 / 100	24 / 80	18 / 60	12 / 40	9 / 30	6 / 20	3 / 10
78	35	0.00	0.00	0.25	0.80	1.40	2.10	3.95	4.50	5.60	7.00	8.10	10.10	13.90	15.25	17.40	20.50	−1.00	−1.00	−1.00
67	30	0.00	0.00	0.25	0.90	1.30	2.20	4.00	4.80	6.00	7.50	8.90	11.00	15.00	16.40	19.00	−1.00	−1.00	−1.00	−1.00
56	25	0.00	0.00	0.40	1.00	1.80	2.50	4.50	5.30	6.75	8.25	9.90	12.00	16.00	−1.00	−1.00	−1.00	−1.00	−1.00	−1.00
45	20	0.00	0.00	0.50	1.00	1.80	2.50	5.00	5.80	7.10	9.10	10.80	13.80	−1.00	−1.00	−1.00	−1.00	−1.00	−1.00	−1.00
34	15	0.00	0.25	0.75	1.20	2.00	3.25	5.50	6.50	8.00	10.10	12.00	−1.00	−1.00	−1.00	−1.00	−1.00	−1.00	−1.00	−1.00
22	10	0.00	0.25	0.50	1.50	2.00	3.10	5.75	6.80	8.60	−1.00	−1.00	−1.00	−1.00	−1.00	−1.00	−1.00	−1.00	−1.00	−1.00
11	5	0.00	0.60	1.00	2.00	2.75	3.75	6.00	−1.00	−1.00	−1.00	−1.00	−1.00	−1.00	−1.00	−1.00	−1.00	−1.00	−1.00	−1.00

Scale Used with Table Above for Determining Soil Loss E_4

Soil Loss $E_3 = I\ K\ C$ or $E_4 = I\ K\ C\ f(L)$ (tons/acre/year and tons/ha/year)

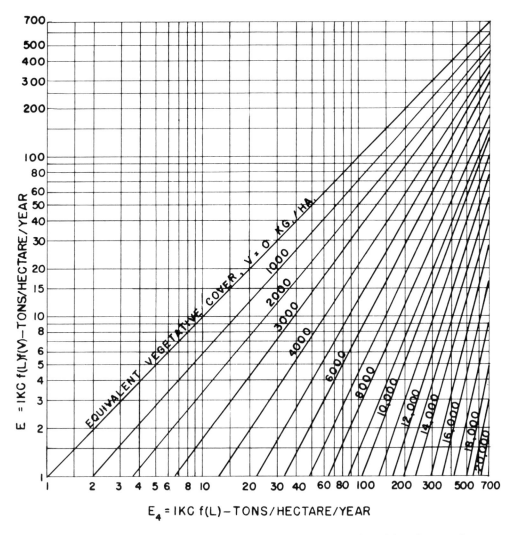

FIGURE 4.   Chart to determine soil loss E = IKCLV from soil loss E_4 = IKCL and from the vegetative cover factor, V.

REFERENCES

1. **Chepil, W. S.,** Sedimentary characteristics of duststorms. III. Composition of suspended dust, *Am. J. Sci.,* 255, 206, 1957.
2. **Gillette, D. A. and Walker, T. R.,** Characteristics of airborne particles produced by wind erosion of sandy soil, high plains of West Texas, *Soil Sci.,* 123, 97, 1977.
3. **Gillette, D. A.,** Fine-particle emissions due to wind erosion, *Trans. ASAE,* 20(5), 890, 1977.
4. **Chepil, W. S.,** Dynamics of wind erosion. I. Nature of movement of soil by wind, *Soil Sci.,* 60(4), 305, 1945.
5. **Gillette, D. A.,** A wind tunnel simulation of the erosion of soil: effect of soil texture, sandblasting, windspeed, and soil consolidation on dust production, *Atmos. Environ.,* 12, 1735, 1978.
6. **Hagen, L. J. and Skidmore, E. L.,** Wind erosion and visibility problems, *Trans. ASAE,* 20(5), 898, 1977.
7. **Smith, W. H.,** Removal of atmospheric particulates by urban vegetation: implication for human and vegetative health, *Yale J. Biol. Med.,* 50(2), 185, 1977.

8. **Free, E. E.**, The Movement of Soil Material by Wind, Bur. Soils Bull. No. 68, U.S. Department of Agriculture, Washington, D.C., 1911.

9. **Bagnold, R. A.**, *The Physics of Blown Sand and Desert Dunes*, Methuen, London, 1941.

10. **Zingg, A. W.**, Wind-Tunnel Studies of the Movement of Sedimentary Materials, Bull. 34, Proc. 5th Hyd. Conf., State University of Iowa Studies in Engineering, Ames, 1953, 111.

11. **White, B. R. and Schulz, J. C.**, Magnus effect in saltation, *J. Fluid Mech.*, 81(3), 497, 1977.

12. **Horikawa, K. and Shen, H. W.**, Sand Movement by Wind Action, Beach Eros. Bull., Corps Eng. Tech. Memo. No. 119, U.S. Army Corps of Engineers, Washington, D.C., 1960.

13. **Woodruff, N. P., Lyles, L., Siddoway, F. H., and Fryrear, D. W.**, How to Control Wind Erosion (revised), USDA Agric. Info. Bull. No. 354, U.S. Department of Agriculture, Washington, D.C., 1977.

14. **Woodruff, N. P. and Siddoway, F. H.**, A wind erosion equation, *Soil Sci. Soc. Am. Proc.*, 29, 602, 1965.

15. **Chepil, W. S., Siddoway, F. H., and Armbrust, D. V.**, Climatic factor for estimating wind erodibility of farm fields, *J. Soil Water Conserv.*, 17(4), 162, 1962.

16. **Skidmore, E. L.**, Assessing wind erosion forces: directions and relative magnitudes, *Soil Sci. Soc. Am. Proc.*, 29(5), 587, 1965.

17. **Skidmore, E. L. and Woodruff, N. P.**, Wind Erosion Forces in the United States and Their Use in Predicting Soil Loss, Agriculture Handbook No. 346, Agricultural Research Service, U.S. Department of Agriculture, Washington, D.C. 1968.

18. **Siddoway, F. H., Chepil, W. S., and Armbrust, D. V.**, Effect of kind, amount, and placement of residue on wind erosion control, *Trans. ASAE*, 8(3), 327, 1965.

19. **Woodruff, N. P., Lyles, L., Dickerson, J. D., and Armbrust, D. V.**, Using cattle feedlot manure to control wind erosion, *J. Soil Water Conserv.*, 29, 127, 1974.

20. **Lyles, L. and Allison, B. E.**, Equivalent Wind Erosion Protection of Selected Crop Residues, Wind Erosion Research Unit, SEA-AR, U.S. Department of Agriculture, Manhattan, Kan.

21. **Skidmore, E. L., Fisher, P. S., and Woodruff, N. P.**, Wind erosion equation: computer solution and application, *Soil Sci. Soc. Am. Proc.*, 34, 931, 1970.

22. **Fisher, P. S. and Skidmore, E. L.**, WEROS: A Fortran IV Program to Solve the Wind Erosion Equation, ARS 41-174, U.S. Department of Agriculture, Washington, D.C., 1970.

23. **Hagen, L. J., Skidmore, E. L., and Dickerson, J. D.**, Designing narrow strip barrier systems to control wind erosion, *J. Soil Water Conserv.*, 27, 269, 1972.

24. PEDCO-Environmental Specialists, Inc., Investigations of Fugitive Dust-Sources, Emissions, and Control, Report prepared under Contract No. 68-02-0044, Task order No. 9, Office of Air Quality Planning and Standards, U.S. Environmental Protection Agency, Research Triangle Park, N.C., 1973.

25. **Wilson, L.**, Application of the wind erosion equation to predict fugitive dust emissions, *J. Soil Water Conserv.*, 30, 215, 1975.

26. **Gillette, D. A., Blifford, I. H., Jr., and Fenster, C. R.**, Measurements of aerosol size distribution and vertical fluxes of aerosols on land subject to wind erosion, *J. Appl. Meteorol.*, 11, 977, 1972.

27. **Lyles, L.**, Possible effects of wind erosion on soil productivity, *J. Soil Water Conserv.*, 30, 279, 1975.

28. **Lyles, L.**, Wind erosion: processes and effect on soil productivity, *Trans. ASAE*, 20(5), 880, 1977.

29. **Lyles, L., Schmeidler, N. F., and Woodruff, N. P.**, Stubble Requirements in Field Strips to Trap Windblown Soils, Kansas Agric. Exp. Stn. Res. Publ. 164, Kansas Agricultural Experiment Station, Manhattan, 1973.

30. **Skidmore, E. L., Kumar, M., and Larson, W. E.**, Crop residue management for wind erosion control in the Great Plains, *J. Soil Water Conserv.*, 34, 90, 1979.

31. **Skidmore, E. L.**, Soil and water management and conservation: wind erosion, in *CRC Handbook Series in Agriculture, Section A: Soils and Climate*, CRC Press, Boca Raton, Fla., 1980.

32. **Lyles, L. and Allison, B. E.**, Range grasses and their small grain equivalents for wind erosion control, *J. Range Manage.*, 33(2), 143, 1980.

ENVIRONMENTAL REQUIREMENTS*

James M. Steichen

INTRODUCTION

Traditionally the adoption of soil conservation practices was based on preserving the soil resource. Land, in our personal care for only a few years, is ultimately passed on to future generations. Thus, if the future population of the world will be able to feed itself, the soil resource must outlive not only past and present generations but future generations as well. Soil erosion in some cases has had significant impact on the soil resource as recorded by Lowdermilk[1] and Carter and Dale.[2] They show how rich and powerful civilizations declined after their natural resources including soil were wasted away through ignorance, greed, and concern with only the present.

As American leaders began to gain knowledge and understanding of the importance of preserving available resources for future generations, they developed a conservation attitude for preservation of resources. This philosophy has formed the basis of support for soil conservation programs in the U.S. for decades. Federal support in the form of technical assistance and cost sharing, slow to come, received a great boost during the Dust Bowl days of the 1930s. These programs, started during a time of world-wide economic problems, have since been continued.

The environmental movement developed renewed strength during the 1960s and early 1970s, and important laws were passed as a result. The Federal Water Pollution Control Act Amendments of 1972 (Public Law 92-500) mandated a sweeping federal-state-community campaign to prevent, reduce, and eliminate water pollution. Section 208 required states to identify nonpoint pollution problems from agriculture and evaluate alternatives for eliminating or reducing the problems. Funding to implement the 208 plans was authorized by the Clean Water Act of 1977. The goals and political support for this Rural Clean Water Program differ from the traditional soil conservation effort. Although many of the practices will be similar, the primary objective of the Rural Clean Water Program is clean water, while primary goal of the traditional soil conservation program is protecting the soil resource.

PROTECTION OF SOIL RESOURCE

Value of Topsoil

Erosion is a natural geological process by which watersheds are formed, mountains worn down, and lowlands filled. Excessive erosion of the productive topsoil by man's activities is our major concern. How much soil can we lose without the loss being excessive?

In general, there are two categories of economic loss due to soil erosion. First, sediment from cultivated land removes plant nutrients and degrades important soil structural characteristics such as permeability, water holding capacity, and tilth. The second loss is deposition of sediment and associated nutrients and chemicals as a pollutant, which is discussed later.

Beasley[3] estimated that the 2.7 billion t of soil eroded from U.S. farms and forest land each year has an average analysis of 0.10% nitrogen (N), 0.15% phosphorus (P_2O_5), and 1.5% potassium (K_2O). This means that 45 million t of plant nutrients are lost by erosion each year in the U.S.

Soil erosion has an important impact on soil properties that affect crop productivity, but

* Contribution 81-201-A, Department of Agricultural Engineering, Kansas Agricultural Experiment Station, Manhattan, Kansas.

that impact is difficult to generalize. Subsoil is almost always lower in organic matter and less permeable than topsoil. As topsoil is eroded, the subsoil absorbs rainfall less rapidly, so that runoff increases and less water is available for crop production. In most cases the subsoil has poorer structure, lower organic matter, and less nutrients. It is more difficult to prepare a seedbed in subsoil, and crop germination and yields are adversely affected.[3]

Pimental et al.[4] estimate that corn yields are reduced an average of about 100 kg/ha for each centimeter of topsoil lost from a basis of 30 cm of topsoil or less; oat yields are reduced by an average of 34 kg/ha; wheat yields, 42 kg/ha; and soybean yields, by 69 kg/ha. The primary reasons for the reduced yields on eroded soils are low nitrogen content, impaired soil structure, deficient organic matter, and reduced availability of stored soil moisture.

While soil is being eroded, new soil is being produced. The rate varies widely depending on soil minerals, weather, etc., but it is generally much slower than the erosion process. Under ideal conditions, Hudson[5] estimates that a centimeter of topsoil may be formed in about 12 years. Other estimates range from 40 to 400 years.[4]

Conservation Decisions and Economics

The economic effects of soil erosion can be relatively minor for a single year and are illustrated with corn. Assuming a reduction of 100 kg/ha in yield per centimeter of topsoil lost and about 45 t/hectare lost annually in continuous corn production, the annual per hectare reduction in yield (from land with 30 cm topsoil or less) would be about 30 kg of corn (worth about $4). A single year's loss then is less than 1%[4] and cannot be measured because of production variabilities.

If soil erosion is still a problem, why have yields continued to increase? Since about 1950, agricultural technologies (primarily the availability and use of cheap fossil fuel in the form of fertilizer, pesticides, new varieties, and more efficient equipment) have sharply increased yields. The massive inputs of those energy factors have easily masked the negative effects of soil erosion. By the late 1970s the rate of production increase was beginning to level out as available ways to substitute energy for soil resources had been effected.

Although the short-term consequences of soil erosion appear minor, the long-term effects are cumulative and considerable, especially in the new era of expensive and increasingly scarce fossil fuel to substitute for lost soil resources. After topsoil is lost, it cannot be replaced.

Fossil fuels as fertilizers and pesticides have been used to offset the loss of productivity. Pimental et al.[4] estimate that the productivity potential of all U.S. cropland has been reduced 10 to 15% by soil loss. The input of fossil energy for crop production is about 7.4 million kcal/ha. About half is used to increase productivity and half to reduce labor. Thus the estimated per hectare input required to offset past soil losses is 490,000 kcal/ha. An estimated 47 ℓ equivalents of fuel per hectare is required annually to offset soil erosion loss on cropland.[4]

When an economic system operates with high interest rates, it requires an immediate return to investment and discounts future benefits. Most conservation practices now require an investment, while most of the benefits are delayed into the future. Inflation and high interest rates lead a manager to make investments that have immediate returns and discourage long-term investment with most benefits in the more distant future. The economics of natural resources are less well developed than many areas of economics. Present means of economic analysis encourage exploitation of natural resources as if they were in unlimited supply. However, reduced petroleum supplies have shown that is not true. While the situation with prime cropland may not be as dramatic, it follows the same principle.

T-Values

The term "soil loss tolerance" (T-value) denotes the maximum erosion that will permit a high crop productivity to be sustained economically and indefinitely. Establishing tolerances for specific soils and topography has been largely a matter of collective judgment, considering both physical and economic factors. In the U.S., maximum soil loss rates thus determined range from 2.2 to 11.2 t/ha/year depending upon soil properties, soil depth, topography, and prior erosion. A deep, medium-textured, moderately permeable soil that has subsoil characteristics favorable for plant growth has a tolerance of 11.2 t/ha/year. Such soils with less favorable subsoil would have a tolerance of 9 t/ha/year. Shallower soils have even less tolerance.[3]

Soil-loss tolerances are established to protect the soil resource, but clean-water requirements may demand soil loss limits based on other factors. Past tolerances have not reflected erosion control required to preserve or restore clean water. Such factors include distance of the field from a major waterway, sediment-transporting characteristics of the intervening area, sediment composition, needs of the particular body of water being protected, and probable fluctuations in sediment loads.[7] Information about specific soil loss tolerances can be obtained from the U.S. Soil Conservation Service.

CLEAN WATER REQUIREMENTS

Sediment

Sediment yield depends on the rate of total erosion in the contributing basin and the transport efficiency of the eroded materials. Both erosion and transport factors vary so widely that any statements concerning sediment yield for a geographic region must be generalized with rather wide limits.[8] Brune[9] portrays the sediment yield variation in selected physiographic areas of the U.S. (Figure 1). Observe the wide variation not only between regions but also within the same region. Natural differences between regions must be recognized in establishing water quality goals.

Only a portion of the soil eroded on a hillside is eventually discharged into the ocean. Most of it is deposited at intermediate points where the runoff cannot sustain transport. Examples are bases of eroding slopes, minor depressions, flood plains, grassed filter strips, reservoirs, and stream channels.

Engineers needing to predict the amount and location of sediment movement have defined sediment delivery ratio as the ratio of sediment delivered at a location in the stream system to the gross erosion from the drainage area above that point.[7] Many factors affect the sediment delivery ratio. No general equation for sediment delivery ratios exists, but several observed relationships provide guidelines for approximating them. The size of the drainage area is important because sediment transported farther downstream has more opportunities to be deposited. Figure 2 summarizes sediment delivery ratios measured in the central and eastern U.S.[8] Limited data have shown that sediment delivery ratios appear to vary inversely as the 0.2 power of the drainage area.[8]

This approach of defining sediment delivery is useful for engineering applications such as designing sediment storage for a reservoir. However, it is not nearly as useful for defining sediment delivery for water quality. From the standpoint of managing water quality at the top of the watershed, the initial delivery of sediment to the drainage system is most important. In such cases, drainage basins are much smaller and fewer data exist than for the earlier definition.

McElroy et al.[10] developed a relationship between sediment delivery ratio and drainage density for relatively homogeneous basins (Figure 3). The horizontal scale of the figure is the reciprocal of drainage density, which is defined as the ratio of total channel segment lengths (accumulated for all orders within a basin) to the basin area. The reciprocal of drainage density may be thought of as an expression of how close channels or natural water courses are spaced or the average distance soil particles travel from an erosion site to the receptor water.[10]

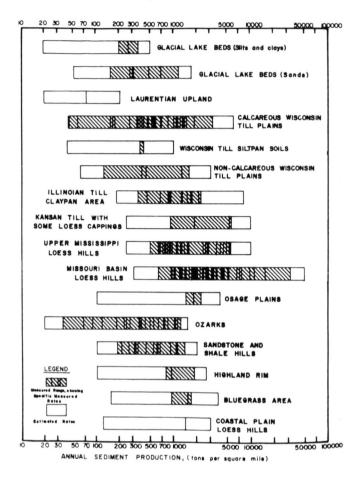

FIGURE 1.   Sediment yields in the U.S., by physiographic areas, in tons per square mile per year.[9]

The relationship in Figure 3 also takes soil texture into account. Clays and silty clays have higher sediment delivery ratios than coarser soils such as sands. The coarser sediments drop from suspension first, while the finer sediments are carried farther downstream. Additional data need to be collected to validate this relationship, and other factors in the deposition mechanism need to be included. Typical values for drainage density for various locales in the U.S. are given in Table 1.[10]

The gross erosion in the basin is the sum of sheet and rill erosion, gully erosion, channel erosion, and other sources. Generally, sheet and rill erosion contribute most of soil losses in a basin. Sheet and rill erosion can be estimated by the Universal Soil Loss Equation,[6] but estimating soil losses from other sources must be based on different methods.[6]

The river carrying a load can be considered a transporting machine that, like all other machines, can be characterized by the relation:

$$\text{rate of doing work} = \text{available power} \times \text{efficiency}^{11}$$

The available power in a river derives from the movement of a mass of water from a higher to a lower elevation. As the potential energy of position is converted to kinetic energy, a major portion of the energy is turned into heat by friction associated with the motion of flow. A small part of the energy is used to transport sediment or alter the channel. The

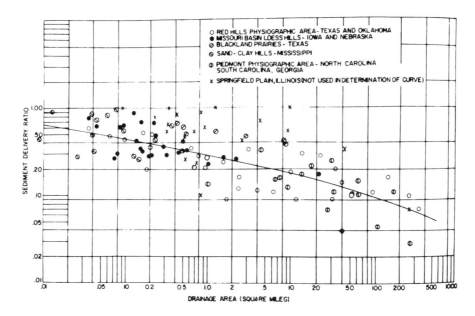

FIGURE 2. Relationship between size of drainage basin and sediment delivery ratio. (From Vanoni, V. A., *Sedimentation Engineering,* ASCE M & R No. 54, American Society of Civil Engineers, New York, 1975. With permission.)

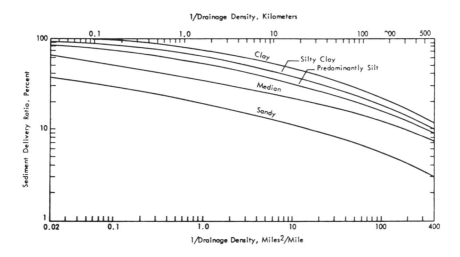

FIGURE 3. Sediment delivery ratio for relatively homogeneous basins.[10]

available power over a unit area of stream bed is proportional to the product of the discharge rate times the slope of the water surface.[11]

When the amount of sediment introduced into a stream exceeds what the stream can transport, the excess must be deposited, and the stream bed is thus built up or aggraded. Conversely, if the rate sediment introduced into the stream is less than the transport capacity, and the channel bed and banks are erodible, the stream will erode or degrade the bed and banks to supply the deficiency.[8]

The major dimensions of aggrading or degrading channels remain in a constant state of change until, by one means or another, equilibrium is established between the sediment inflow and discharge.[8]

Table 1
TYPICAL DRAINAGE DENSITIES

Location	Drainage density	
	km/km²	mi/mi²
Appalachian plateau province	1.9—2.5	3.0—4.0
Central and eastern U.S.	5—10	8.0—16.0
Dry areas of the Rocky Mountain region	31—62	50—100
The Rocky Mountain region (except dry areas)	5—10	8.0—16
Coastal ranges of Southern California	12—25	20—40
Badlands in South Dakota	125—250	200—400
Badlands in New Jersey	183—510	310—820

Nutrients

Eutrophication is the process by which lakes age. It is a natural process, just as erosion is a natural process. Cultural eutrophication speeding the process is the problem, just as accelerated soil erosion is a problem. Enriching lake water with nutrients from runoff rich in nitrogen and phosphorus encourages growth of aquatic vegetation. Rapid growth of algae is the major eutrophication problem in most states. Excessive algae growth can cause taste and odor problems for drinking water supplies, and when a large algae "bloom" dies and decays, dissolved oxygen is consumed so fast that fish may suffocate.

Nitrate is highly soluble and moves with the water. Sediment, on the other hand, is the major transport vehicle for phosphorus and organic nitrogen, so that control of sediment can reduce loss of phosphorus and organic matter but is not as likely to affect nitrates, which move with runoff water.

Pesticides

The pathway by which a pesticide enters surface water is governed by sorption and solution equilibria that depend primarily on water solubility, the degree and strength of adsorption on soil particles, and on the interaction of both soil and pesticides with water.[12] Generally, compounds that are more water-soluble will move as dissolved in runoff water, and those more strongly adsorbed will move mostly on sediment.

Allowable levels of pesticide contamination for public water supplies, set by the National Safe Drinking Water Act, illustrate the low concentrations allowed: endrin, 0.0002 mg/ℓ; lindane, 0.004 mg/ℓ; and 2,4-D, 0.1 mg/ℓ.

LEGAL REQUIREMENTS

Wind Erosion

During the dark days of the "Dirty 30s", states in the Great Plains passed laws requiring conservation practices to control wind erosion. Kansas, for example, passed a "Wind Blown Dust and Soil Erosion" law[13] in 1937, which states that "soil erosion caused by wind and dust storms produced thereby is recognized and declared to be destructive of the natural resources of the state and a menace to the health and well being of its citizens." The law made it the duty of landowners to reduce or prevent dust blowing from their land. County commissioners, finding a significant wind erosion problem, may order a landowner to perform emergency tillage to control wind erosion, and after reasonable time hire the tillage done, with its cost a special assessment against the land.

Although now used rarely, the wind erosion laws still exist. They emphasized emergency short-term solutions, while long-term solution by prevention has generally been through voluntary acceptance of practices such as stubble mulching. Interestingly, water erosion laws are considered new and innovative while wind erosion laws have existed for several decades.

Water Erosion

Present Approach

Soil and Water Conservation District laws passed by the states in the 1930s, with the philosophical objective to protect the soil, brought opportunities for voluntary soil conservation.

By the 1960s, soil disturbing activities, especially in quickly urbanizing areas, brought major erosion and flooding problems. Construction, land development, and mining all contributed. As a result, some states and local governments established mandatory requirements to reduce sediment losses, but usually exempted agricultural and silvicultural land. Compliance usually required that an acceptable erosion control plan be developed to obtain a building permit.

In 1971, Iowa passed the first state law setting mandatory soil loss limits on all land.[14,15] This law declares accelerated soil erosion a nuisance and requires abatement of such nuisance when a complaint is filed with the commission of a conservation district. Walker and Cox[16] have reviewed applicable laws related to soil erosion and sedimentation in other states.

Program Guidelines

An idealized outline for developing program guidelines is presented. Recognizing that soil conservation and water quality laws and programs are in a state of evolution, an approach for determining alternative solutions, rather than a particular solution, is presented.

The first step is to determine objectives, which may be either water quality- or soil resource-related or both. Be as specific as possible. Identity specific land uses and soil types with major problems. Water quality goals should identify specific pollutants and beneficial uses affected by the pollutant.

Since developing a program is more than an academic or technical process, the public should be involved at an early stage. Successfully implementing a program requires support from the public most affected by it. If farmer-landowners and other interested people help identify objectives, they will more likely support the means to achieve the objectives.

After identifying the specific problems, identify the alternative practices that technologically can reduce or solve the problem. The U.S. Environmental Protection Agency defines Best Management Practices (BMPs) as " . . . a practice or combination of practices that is determined by a designated planning agency after problem assessment, examination of alternative practices and appropriate public participation to be the most effective practicable (including technological, economic and institutional considerations) means of preventing or reducing the amount of pollution generated by nonpoint sources to a level compatible with water quality goals."[17] Thus, by definition, a BMP must not only reduce pollution, but also meet economic and political restraints.

The most challenging step is developing and adopting a program to implement the BMPs. A trade-off between crop production and water quality must be made. What level of water quality is required by law? There is uncertainty about the water quality improvement that soil conservation practices can provide. Several billion dollars may be required to complete such a program in a single state. Is it possible to provide the investment necessary in light of other needs? A solution determined through the political process will take into account beliefs and values held by society as well as facts determined by research.

The final step is realizing that BMPs are a continuing process. Additional experience, research, and program results likely will require revising of water quality goals, BMPs, or means of implementing the program.

Conclusions

There is a natural tendency to select soil conservation practices to implement a water quality program. This is fine if the identified problem is sediment, but when the major problem is pesticides or other pollutants, soil conservation practices might not provide the solution. Practices dealing with time and amounts and means of pesticide application likely would be more useful.

Because water quality depends heavily on the geology of the region (Figure 1), water quality goals will vary considerably by region. The use of the stream also will affect water quality goals.

Adopting some BMPs may have unexpected results. For example, in a semiarid region, such soil conservation practices as level terraces, ponds, and stubble mulching can effectively reduce soil erosion and reduce runoff. In a semiarid region, reducing runoff by 2 or 3 cm a year may reduce water yield by half or more and thus deplete the resource it was supposed to improve.

Solving some of our environmental problems and maintaining resource productivity may take new and innovative planning, which, to succeed, may involve political and social innovations.

REFERENCES

1. **Lowdermilk, W. C.,** Conquest of the Land Through Seven Thousand Years, Agric. Inf. Bull. No. 99, U.S. Department of Agriculture, Washington, D.C., 1953.
2. **Carter, V. S. and Dale, T.,** *Topsoil and Civilization,* University of Oklahoma Press, Norman, 1974.
3. **Beasley, R. P.,** *Erosion and Sediment Pollution Control,* Iowa State University Press, Ames, 1972.
4. **Pimental, D., Terhune, E. C., Dyson-Hudson, R., Rochereau, S., Samis, R., Smith, E. A., Denman, D., Reifschneider, D., and Shepard, M.,** Land degradation: effects on food and energy resources, *Science,* p. 149, 1976.
5. **Hudson, N.,** *Soil Conservation,* BT Batsford Ltd., London, 1971.
6. **Wischmeier, W. H. and Smith, D. D.,** Predicting Rainfall Erosion Losses — A Guide to Conservation Planning, USDA Agric. Handbook No. 537, U.S. Department of Agriculture, Washington, D.C., 1978.
7. **Stewart, B. A., Woolhiser, D. A., Wischmeier, W. H., Caro, J. H., and Frere, M. H.,** Control of Water Pollution from Cropland, Vols. 1 and 2, Rep. No. EPA-600/2-75-026, U.S. Environmental Protection Agency, Washington, D.C., 1975.
8. **Vanoni, V. A.,** *Sedimentation Engineering,* ASCE M & R No. 54, American Society of Civil Engineers, New York, 1975.
9. **Brune, G. B.,** Collection of Basin Data on Sedimentation, Soil Conservation Service, U.S. Department of Agriculture, Milwaukee, 1953.
10. **McElroy, A. D., Chiu, S. Y., Nebgen, J. W., Aleti, A., and Bennett, F. W.,** Loading Functions for Assessment of Water Pollution from Nonpoint Sources, No. EPA-600/2-76-151, U.S. Environmental Protection Agency, Washington, D.C., 1976.
11. **Dunne, T. and Leopold, L. B.,** *Water In Environmental Planning,* W. H. Freeman & Co., San Francisco, 1978.
12. **Spencer, W. F.,** Distribution of pesticides between soil, water, and air, in Pesticides in the Soil: Ecology, Degradation, and Movement, Int. Symp. Pesticides in the Soil, Michigan State University, East Lansing, 1970. 120.
13. Wind Blown Dust and Soil Erosion, Kansas Statutes Ann. ch. 2, art. 20, 1975.
14. **Fletcher, D. A.,** *Erosion and Sediment Control Programs: Six Case Studies,* National Association of Conservation Districts, League City, Tex., 1976.
15. **Lindquist, D.,** The Iowa program, *Agric. Eng.,* 56(2), 14, 1975.
16. **Walker, W. R. and Cox, W. E.,** Soil erosion and sedimentation: analysis of applicable law, *Trans. ASAE,* 21(2), 282, 1978.
17. Setting the Course for Clean Water, National Wildlife Federation, Washington, D.C., 1977.

Drainage Engineering

LAND GRADING

G. O. Schwab

INTRODUCTION

Land grading for improving the movement of water on the land surface is applicable for irrigation as well as for drainage. Grading is also called land forming, land shaping, or land leveling. Land smoothing is the practice of removing the minor surface irregularities after land grading or leveling. Recommended ASAE terminology for drainage was published by Phillips.[1] For basin or level irrigation, the surface is made truly level. For furrow or flood irrigation and for surface drainage, some minimum grade is desirable. Land grading may be useful on land with considerable slope, such as the area between parallel terraces or on undulating land with depressions or potholes. Generally, land grading applies to relatively flat land where surface water movement is slow. Land grading for both irrigation and drainage is discussed by Phelan,[2] Phillips,[3] and Schwab et al.[4] Gain[5] describes the nature and scope of surface drainage in eastern North America. Grading for drainage was reported by Coote and Zwerman,[6] Beer and Shrader,[7] Saveson,[8] and Walker and Lillard.[9]

Slopes, cuts, and fills in land grading are influenced by soil type, location and elevation of the water source and drainage outlet, climate, crops, and method of irrigation. Soils with shallow topsoil present a major limitation to land grading because exposed subsoil results in reduced plant growth. Topsoil may be stockpiled and redistributed over the cut area, but costs are greatly increased. Reduced crop growth may occur on fill areas, although exposure of subsoil, such as claypan, hardpan, rock, or gravel, is usually more serious. A uniform design slope is more essential for irrigation than for drainage. A variable slope as for drainage will permit less cut and fill than a uniform slope. Accuracy of grading is important for crops that are sensitive to excesses or shortages of water. For flood irrigation, slopes in both directions may be restrictive, whereas for furrow irrigation, the furrow grade largely determines the slope. Land grading can be made compatible for both drainage and irrigation. With irrigation, the largest flow occurs at the upper end of the slope, while for drainage, flow accumulates uniformly along the slope with the highest runoff at the lower end.

METHODS OF GRADING

Several methods for computing design slopes from which cuts and fills can be computed will be presented. Numerical examples can be found in Schwab et al.[4] Field data should be obtained from an instrument survey using a 30-m (100-ft) square grid with elevations taken to the nearest 1 cm (or 0.1 ft). Elevations are also taken at the water supply ditch, drainage outlet, and other critical control points. Laser-equipped machines may also provide the needed grid elevations.

Plane Method

This method assumes that the field is to be graded to a true plane. The average elevation of the field is determined, and this elevation is assigned to the centroid of the area. The centroid is located by taking moments about two perpendicular reference lines. This procedure can be simplified by locating the grid system so that each grid point is at the center of the grid square. Each grid point thus represents the same area. Any plane passing through the centroid will produce equal volumes of cut and fill. The general equation for a plane surface is

$$E = a + S_x X + S_y Y \tag{1}$$

where E = elevation at any point, a = elevation at the origin, and S_x and S_y = slope in the X and Y directions, respectively. The slope of any line in the X or Y direction on the plane of best fit can be determined by the statistical least-squares procedure presented by Chugg.[10] The least-squares plane by definition is that which gives the smallest sum of all the squared differences in elevation between the grid points and the plane. This method does not necessarily provide the best slope for irrigation. For rectangular fields the slope in the X direction is

$$S_x = \frac{\Sigma(XE) - nX_cE_c}{\Sigma X^2 - nX_c^2} \tag{2}$$

where n = total number of grid points, X_c = X distance to the centroid, E = elevation of a grid point, and E_c = elevation of the centroid (average elevation of all points). The slope S_y can be obtained from Equation 2 by substituting Y for X. Although Equation 2 is valid only for rectangular fields, a satisfactory solution can often be obtained by taking one or more arbitrarily selected rectangular areas within the field and extending the slopes of the plane to the remaining areas. Since the plane of best fit must pass through the centroid, substituting S_x and S_y in Equation 1 will give the elevation of the origin. Elevations at each grid point are then computed from Equation 1.

Because of the many calculations to be made, digital computers are a valuable tool for solving such problems. Programs have been developed, such as those presented by Benedict et al.[11] and Smerdon et al.[12] In addition to minimizing cuts and fills, the print-out gives both the least total volume of earth moved and the least distance transported. Raju[13] proposed a so-called fixed-volume center method for computing the slope of a plane which is somewhat simpler, but it is not generally accepted in the U.S. Other methods to be described are simpler than the plane method. These are also easily adapted to frequent changes in slope and different slope lengths.

Profile Method

With this method, ground profiles are plotted and a grade is established that will provide an approximate balance between cuts and fills as well as reducing haul distances to reasonable limits. Cuts and fills are estimated by trial and error and determined graphically from the original and design profiles. If needed, profiles can be plotted at right angles to the original lines. In this way, proposed grade lines can be adjusted to meet cross-slope and downgrade criteria.

By plotting adjacent profiles next to each other on a graph, the design grades of one profile can be easily compared with others. This method is commonly used, especially for small fields.

Plan-Inspection Method

With this method, the grid point elevations are recorded on the plan, and the design grade elevations are determined by inspection after a careful study of the topography. It is largely a trial-and-error procedure, keeping in mind downgrade and cross-slope limitations. As will be explained later, the desired cut-fill ratio and volumes of earthwork are estimated from the summation of all cuts and fills. Although this method does not assure minimum cuts and fills or the shortest length of haul, it is a rapid method, especially for the experienced designer.

Contour-Adjustment Method

To apply this method, a contour map is drawn, and the proposed ground surface is shown on the same map by drawing in new contour lines. The uniformity of slope is controlled by properly spacing the new contours. As with the plan-inspection method, it is largely a trial-and-error procedure. The proper balance between cuts and fills is estimated graphically at the grid points by interpolating between contour lines and by taking the difference between the original and the new surface.

EARTHWORK VOLUME

Several methods are suitable for computing cut or fill volume from a grid survey. The total volume is the summation of the computed volumes of all grid squares plus any irregular grids around the edges of the field. A common procedure, called the four-point method by the U.S. Soil Conservation Service,[14] is sufficiently accurate for land grading. Volume of cuts for each grid square are

$$V_c = \frac{L^2(\Sigma C)^2}{4(\Sigma C + \Sigma F)} \tag{3}$$

where V_c = volume of cut (L^3), L = grid spacing (L), C = cut on the grid corners (L), and F = fill on the grid corners (L).

For computing the volume of fills V_f, $(\Sigma C)^2$ in the numerator of Equation 3 is replaced by $(\Sigma F)^2$. If the volume for an irregular-shaped grid or for only a portion of a grid square is desired, the full grid volume is reduced in proportion to the reduced area of the grid. Another more approximate method is to assume that the cut or fill at a grid point represents the average for half the grid distance from the point in four directions. This method is only suitable for preliminary estimates.

Experience has shown that, in land grading or leveling, the cut-fill ratio should be greater than 1. Compaction from equipment in the cut area, which reduces the volume, and also compaction in the fill area, which increases the fill volume needed, are believed to be the principal reasons for this effect. Marr[15] stated that on level ground between stakes the operator of manually controlled equipment has an optical illusion of a dip in the middle, and therefore in filling, crowning often occurs. The ratio of cut to fill volume is usually 1.3 to 1.6, but may range from 1.1 to 2.0. With the plane method of computing cuts and fills, a settlement correction for the whole field is more convenient to apply. The settlement allowance or the amount of lowering of the elevation may range from 0.3 to 1 cm for compact soils and from 1.5 to 5 cm for loose soils. A small change in elevation will cause a considerable change in the cut-fill ratio. If extra quantities of soil are needed outside the area to be leveled, such as for a roadway or depression, the plane surface can be lowered by the amount of earthwork required. In computing costs, normally the volume of cut is the basis for computation.

With the four-point method, the volume of cut for the grid in Figure 1 is 24.30 m³ and the volume of fill is 10.80 m³. Cuts and fills are normally taken to the nearest centimeter or to tenths of a foot and so recorded on the map.

Volumes by the average end-area method can be computed from the equation

$$V = L/2 \, (A_1 + A_2) \tag{4}$$

where A_1 = cross-sectional area of cut or fill along the first profile (L^2) and A_2 = cross-sectional area of cut or fill along the adjacent profile (L^2). In Figure 1, the volume of cut with the average end-area equation is 24.60 m³ and the volume of fill is 11.10 m³.

The prismoidal formula is

$$V = (L/6) \, (A_1 + 4 \, A_m + A_2) \tag{5}$$

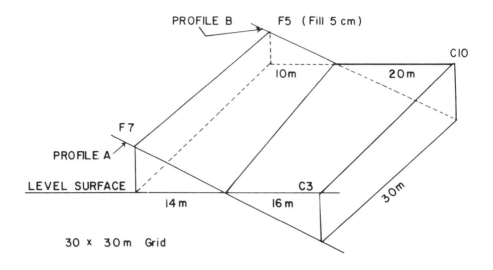

FIGURE 1.   A grid square with both cuts and fills.

Volume Method Eq. (No.)	Volume for Grid in Meters^3	
	Cuts	Fills
4 Point (3)	24.30	10.80
Avg. End Area (4)	24.60	11.10
Prismoidal (5)	24.40	10.90

where A_m = cross-sectional area midway between the end areas A_1 and A_2 (L^2). This method gives the exact volume, but it is seldom used because A_m is cumbersome to compute. The volume of cuts with this method for Figure 1 is 24.40 m³, and the volume of fills is 10.90 m³. The volumes from the four-point method are less than the above, whereas the volumes from the average end-area method are higher. In this example the error is about twice as much for the average end-area method compared to the four-point method. Because of its accuracy and adaptability to digital computers, the four-point method is usually preferred.

If the four points of a grid are all cuts or all fills, the volumes computed are the same for all three methods described above. With the four-point method the computed volume is the same regardless of the number or location of the cuts and fills on the grid corners. With soil prisms more complex than shown in Figure 1, the average end-area or the prismoidal methods are too involved for practical use.

CONSTRUCTION

Prior to making the survey, heavy vegetative growth and residue should be removed. The surface of the soil should be fairly compact and as smooth as possible. Heavy carrier-type scrapers or pans especially designed for accurate depth control at shallow cuts are common. These are powered with crawler or rubber-tired units. Grades should be maintained as accurately as possible. Laser grade-control systems that automatically control the cut or fill depth on the scraper (see Figure 2) are in common use. These systems have also been adapted to dozers and blade graders. Accuracy with laser equipment is much higher than with operator-controlled machines. Cuts and fills are marked on the grid stakes in the field with operator-controlled equipment. With laser units, surveying may be done with the equipment by driving over the field and recording elevations as the laser receiver passes grid intersections

FIGURE 2. Land grading equipment and laser for grade control. (Courtesy of Laserplane Corp., Dayton, Ohio.)

or other identifiable coordinates. Rotating laser units have a range of about 300 m from the transmitter with good accuracy and can be tilted to slopes of about 10%. Where the field slopes in two directions, the direction and magnitude of the maximum slope must be determined to properly set the laser transmitter. This maximum slope is sometimes referred to as the compound slope.

The maximum slope of a field may be determined from Equation 6. As shown in Figure 3, any point (X,Y) on the line of maximum slope may be designated by the coordinates (L sin A, L cos A) where L is the horizontal distance from the origin to the point. Substituting in Equation 1 and dividing by L,

$$S_{(max.)} = E/L = S_x \sin A + S_y \cos A \qquad (6)$$

where A = the angle from the line of maximum slope to the direction of major slope. Differentiating Equation 6 with respect to A and setting equal to zero gives

$$\frac{S_x}{S_y} = \frac{\sin A}{\cos A} = \tan A. \qquad (7)$$

For example, if S_x = 0.03 and S_y 0.04, tan A = 0.75 for which A = 36.87°. Substituting in Equation 6 gives $S_{(max)}$ = 0.05. This slope is greater than that in either the X or Y directions, but is the correct slope to set the laser rotating beam transmitter. Computed slopes and directions for other slopes are given in Table 1. The line of maximum slope is perpendicular to the contour lines. Where grading is done in one direction only, the correct slope for the transmitter is the design slope in that particular direction. If one slope is zero, the correct slope is that for the other direction. For a single plane surface, the slope (maximum) is the same over the entire area.

Where more than one design plane surface has been established in a field, the operator may reset the transmitter for each plane or a grade breaker can be used. The grade breaker will change the detector height as the equipment moves through the field so that the proper

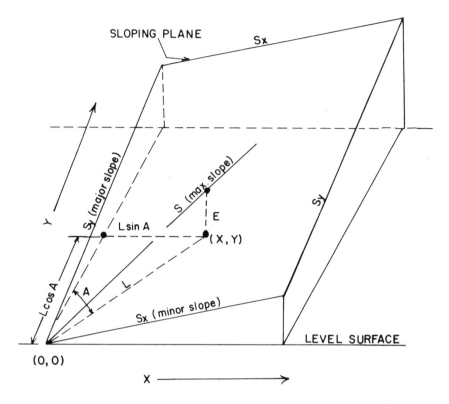

FIGURE 3. Geometry of a field to obtain maximum slope.

Table 1
MAXIMUM SLOPE AND ANGLE FOR FIELDS WITH SLOPES IN TWO DIRECTIONS

Major (larger) slope (%)	Minor (smaller) slope (%)[a]							
	0.01	0.02	0.04	0.06	0.08	0.10	0.20	0.30
0.01	0.014 (45)							
0.02	0.022 (26)	0.028 (45)						
0.04	0.041 (14)	0.045 (26)	0.057 (45)					
0.06	0.061 (9)	0.063 (18)	0.072 (34)	0.085 (45)				
0.08	0.081 (7)	0.082 (14)	0.089 (26)	0.100 (37)	0.113 (45)			
0.10	0.100 (6)	0.102 (11)	0.108 (22)	0.117 (31)	0.128 (39)	0.141 (45)		
0.20	0.200 (3)	0.201 (6)	0.204 (11)	0.209 (17)	0.215 (22)	0.224 (26)	0.283 (45)	
0.30	0.300 (2)	0.301 (4)	0.303 (8)	0.306 (11)	0.310 (15)	0.316 (18)	0.361 (34)	0.424 (45)

[a]   Top value is maximum slope in percent. Direction of slope from major slope in degrees (in parentheses).

From Owner Operator's Manual, Laserplane Corp., Dayton, Ohio, 1980. With permission.

grade is maintained at all times. Manufacturer instructions manuals should be carefully followed.

Common tolerance from the design grade in land grading is about 2 to 3 cm (0.05 to 0.01 ft). Accuracy is more critical for irrigation than for drainage, but no reverse grade is permitted. After the major earthwork has been completed, minor irregularities may be removed by land smoothing using a land leveler or float. Levelers should go over the field in both directions, then diagonally in both directions, and finally in the direction of drainage or irrigation.

MAINTENANCE

After heavy rainfall or irrigation, unequal settlement or subsidence may take place in which the fill areas usually settle more than the cut areas. Upon wetting, cut areas may swell more than fill areas. After a year or so, regrading and smoothing may be required. The same procedures as described for the original construction should be followed. Regular maintenance may be required every few years. It is often difficult to find time for maintenance without interfering with normal farm operations.

REFERENCES

1. **Phillips, R. L.,** Committee report on surface drainage systems, *Trans. ASAE,* 6(4), 313, 1963.
2. **Phelan, J. T.,** Land leveling and grading, in *Agricultural Engineers' Handbook,* Richey, C. B., Jacobson, P., and Hall, C. W., Eds., McGraw-Hill, New York, 1961, chap. 35.
3. **Phillips, R. L.,** Land leveling for drainage and irrigation, *Agric. Eng.,* 39, 463, 1958.
4. **Schwab, G. O., Frevert, R. K., Edminster, T. W., and Barnes, K. K.,** *Soil and Water Conservation Engineering,* 3rd ed., John Wiley & Sons, New York, 1981.
5. **Gain, E. W.,** Nature and scope of surface drainage in United States and Canada, *Trans. ASAE,* 7(2), 167, 1964.
6. **Coote, D. R. and Zwerman, P. J.,** Surface Drainage of Flat Lands in the Eastern United States, Cornell University Ext. Bull. 1224, Cornell University, Ithaca, N.Y., 1970.
7. **Beer, C. E. and Shrader, W. D.,** Response of corn yields to bedding soils, *Agric. Eng.,* 42, 618, 1961.
8. **Saveson, I. L.,** Land forming for drainage, *Agric. Eng.,* 40, 208, 1959.
9. **Walker, P. and Lillard, J. H.,** Land forming in the southeast, *J. Soil Water Conserv.,* 16, 166, 1961.
10. **Chugg, G. E.,** Calculations for land gradation, *Agric. Eng.,* 28, 461, 1947.
11. **Benedict, R. et al.,** Land grading and leveling programs for digital computers, Arkansas Agric. Exp. Stn. Bull. 691, 1964.
12. **Smerdon, E. T., et al.,** Electronic computers for least-cost land forming calculations, *Trans ASAE,* 9, 190, 1966.
13. **Raju, V. S.,** Land grading for irrigation, *Trans. ASAE,* 3(1), 38, 1960.
14. Irrigation, National Engineering Handbook, Section 15, Land Leveling (lithograph), Soil Conservation Service, U.S. Department of Agriculture, Washington, D.C., 1959, chap. 12.
15. **Marr, J. C.,** Grading land for surface irrigation, Calif. Agric. Exp. Circ. 438, 1957.
16. Owner Operator's Manual Laserplane Corp., Dayton, Ohio, 1980.

SURFACE DRAINAGE SYSTEMS

Byron H. Nolte

INTRODUCTION

Surface drainage is the orderly removal of excess water from the surface of the land through grading or shaping of the land surface, constructing ditches, and improving natural channels.

Figure 1 illustrates the general layout of a parallel-surface drainage system and surface shaping. A random-surface drainage system is illustrated in Figure 2.[1] A field has good surface drainage if the depth of surface water storage is 2.5 mm.[2] This degree of surface drainage requires careful tillage operations and land smoothing every few years.

The objective of surface drainage is to remove excess water so that farming operations can be performed in a timely manner, crop yields can be maximized, and soil erosion can be controlled.

DESIGN

Surface drain design criteria have been developed for crop production on flat land, average slope less than 0.5%.

Curves proposed by Sutton,[3] known as Soil Conservation Service[4] (SCS) drainage runoff curves (A, B, C, D curves), have been used in the northern portion of the humid area for 5- to 520-km[2] watersheds. These curves have been extended to 5-ha watersheds by SCS and the American Society of Agricultural Engineers.[5] The curves to estimate runoff may be expressed using the general formula:

$$Q = Cm^x$$

Expressed in SI units the formula is

$$Q = 0.0283 \ C(0.386 \ M)^x$$

where Q = flow in m[3]/sec, C = drainage coefficient from Table 1, M = area in km[2], and x = exponent for the region from Table 1.

Table 1 gives the range of drainage areas, drainage coefficients, and exponents to use in the above formula for various regions of the country. The regions where the coefficients and exponents are applicable are shown in Figure 3.[6]

Field drains are located at the lower end of field rows, through surface depressions, above barriers that trap runoff, and where required to divert runoff from lower areas. Field drains need not be designed to carry the quantities of flow indicated by the drainage curves or design equations, as their primary purpose is to remove residual surface water after volume runoff has passed out of the field.

Recommended humid areas row and field drain dimensions (Table 2) have been developed by the American Society of Agricultural Engineers.[7] Recommended grades are from 0.1 to 0.3%. Field lateral side slope recommendations are shown in Table 3. The outlet should be 0.3 m deeper than the field ditches.

The Bureau of Reclamation recommends that design for surface flows in western irrigated areas be made using the McMath formula.[8,9] Results are considered reliable for planning purposes. The McMath formula is

SHALLOW DITCHES TO INTERCEPT AND REMOVE SURFACE WATER FROM THE
FIELD AND REDUCE THE LENGTH OF ROW DRAINAGE ON FLAT LANDS

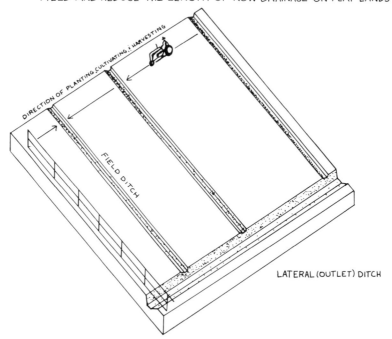

TYPICAL CROSS SECTION OF GROUND SURFACE

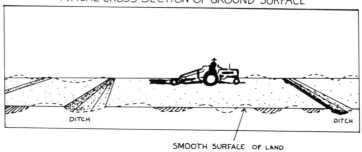

FIGURE 1.   Parallel surface drainage system. (Drawn by Ann M. Nolte.)

$$Q = C \, i \, S^{1/5} \, A^{4/5}$$

where Q = flood discharge, C = coefficient representing the basin characteristics, i = rate of rainfall for the time of concentration and frequency, S = fall of main channel between the farthest contributing point and the point of concentration, and A = area of basin.

Detailed runoff computations may be made for both humid and arid areas using the SCS runoff curve number or other procedure.[10]

Channel dimensions may be determined using the Manning equation to determine the mean velocity (v) and the relation:

$$Q = Av$$

where Q = flow rate in m³/sec, A = cross-sectional area of the channel in m², and v = mean velocity in m/sec.

The Manning's equation is

SMOOTH AREA SO LAND WILL DRAIN TO THE LARGE DEPRESSIONS OR RANDOM DITCHES

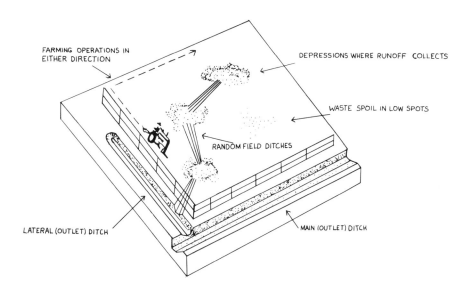

TYPICAL CROSS SECTION OF A RANDOM DITCH

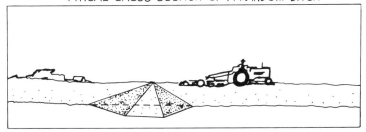

FIGURE 2. Random surface drainage system. (Drawn by Ann M. Nolte.)

Table 1
DRAINAGE COEFFICIENTS AND EXPONENTS FOR VARIOUS AREAS OF THE COUNTRY[5,6]

Location and land use	Area (km²)	C	x
Northern humid area			
Good protection from overflow (A)	5—520	150	0.6
Excellent drainage (B)	0.05—520	72	0.7
Cropland, good drainage (C)	0.05—520	37	0.83
Pasture, fair drainage (D)	0.05—520	25	0.83
Red River Valley, cultivated	0.05—260	20	0.83
Southern humid area			
Southwest maximum hill	2.6—260	131	0.7
Southwest minimum hill	2.6—260	80	0.753
Coastal plain, cultivated	0.05—260	45	0.83
Delta and river bottom, cultivated	0.05—260	40	0.83
Improved pasture	0.05—260	30	0.83
Irrigated rice	0.05—260	22.5	0.83
Unimproved pasture and range	0.05—260	15	0.83
Woodland	0.05—260	10	0.83

FIGURE 3.   Aerial application of drainage coefficients, C, and exponents, x.

Table 2
RECOMMENDED ROW DRAIN AND FIELD DRAIN DIMENSIONS[5]

Type	Depth ft	Depth m	Bottom width ft	Bottom width m	Side slope
Row drain	0.3[a]	0.09[a]	0	0	—
Vee	0.5—1.0	0.15—0.3	0	0	10:1 or flatter[b]
One half vee	0.5—1.0	0.15—0.3	—	—	15:1 or flatter
Trapezoidal	0.75—1.5	0.23—0.45	8	2.4	8:1 or flatter

[a]   Minimum, 0.06 m (0.2 ft) below row middles for row crops.
[b]   Ten horizontal to one vertical.

Table 3
RECOMMENDED FIELD LATERAL SIDE SLOPES[5]

Cross-section type	Depth ft	Depth m	Recommended side slope	Minimum side slope
Vee	1.0—2.0	0.3—0.6	6:1	3:1
Vee	2.1 and over	0.61 and over	4:1	3:1
Trapezoidal	1.0—3.0	0.3—0.9	4:1	2:1
Trapezoidal	3.1 and over	0.91 and over	2.5:1	1:1

Table 4
DITCH DESIGN "N"
VALUES[6]

Hydraulic radius, m	"n"
Less than 0.75	0.040—0.045
0.75—1.25	0.035—0.040
1.25—1.5	0.030—0.035
More than 1.5	0.025—0.030

Table 5
PERMISSIBLE VELOCITIES[4]

Soil texture	Velocity (m/sec)
Sand and fine sandy loam (noncolloidal)	0.8
Silt loam (also high lime clay)	0.9
Sandy clay loam	1.1
Clay loam	1.2
Stiff clay, fine gravel, graded loam or gravel	1.5
Graded silt to cobbles (colloidal)	1.7
Shale, hardpan, and course gravel	1.8

$$v = \frac{r^{2/3} \, s^{1/2}}{n}$$

where v = mean velocity in m/sec, r = hydraulic radius in meters, cross-sectional area of the channel divided by its wetted perimeter, s = slope in m/m, and n = coefficient of roughness, Manning's "n". A guide for selection of design "n" values is given in Table 4.

The maximum permissible velocities for open drains according to soil texture are given in Table 5. In noncohesive soils, the tractive force analysis may be used to determine stability of the drainage channel.[11,12]

The design frequency of drainage channels is based primarily on experience. For large investments, an analysis of benefits, costs, and risks would be appropriate.

CONSTRUCTION

The construction methods and equipment should be considered when designs are selected. Environmental impacts and maintenance plans should be considered when specifications are prepared.[13]

MAINTENANCE

A maintenance plan should be developed along with design and construction specifications. The owner should be provided criteria for determining when maintenance is needed. Channel side slopes determine the equipment applicable for maintenance (Table 6).

OUTLETS

A drainage system is only as good as its outlet. Most systems outlet into a public water course that is the responsibility of some district or unit of government. Plans and designs

Table 6
CHANNEL SIDE SLOPES FOR MAINTENANCE
METHODS AND EQUIPMENT[4]

Type of maintenance	Recommended steepest side slopes[a]	Remarks
Mowing	3:1	Flatter slopes desirable
Grazing	2:1 or flatter	For ditches greater than 4 ft deep
	1½:1 or flatter	For ditches less than 4 ft deep
Dragline	½:1	Usually used on ditches with steep side slopes, greater than 4 ft deep
Blade equipment	3:1	Flatter slopes desirable
Turning plows	3:1	Flatter slopes desirable
Chemicals	Any	Use caution near crops

[a]   Soil stability may require flatter slopes.

should consider the limitations of the outlet as well as downstream consequences of the proposed improvement. Many states have a statutory procedure for implementing improvements of drains that affect more than one owner.

REFERENCES

1. Ohio Drainage Guide, Ohio Agricultural Research and Development Center, Cooperative Extension Service, The Ohio State University, Ohio Department of Natural Resources, Division of Lands and Soil, U.S. Department of Agriculture Soil Conservation Service, Columbus, 1973, reprinted 1976.
2. **Skaggs, R. W.,** *Evaluation of Drainage-Water Table Control Systems Using a Water Management Model,* Proc. 3rd *Natl.* Drainage Symp., American Society of Agricultural Engineers, St. Joseph, Mich., 1976.
3. **Sutton, J. G.,** Hydraulics of open channels, *Agric. Eng.,* 20(5), 175, 1939.
4. Engineering Field Manual, U.S. Soil Conservation Service, Washington, D.C., 1969.
5. Design and Construction of Surface Drainage Systems on Farms in Humid Areas, ASAE EP 302.2, *Agricultural Engineers Yearbook,* American Society of Agricultural Engineers, St. Joseph, Mich., 1979.
6. Drainage of Agricultural Land, National Engineering Handbook, Section 16, U.S. Soil Conservation Service, Washington, D.C., 1971.
7. Design and Construction of Surface Drainage Systems on Farms in Humid Areas, ASAE EP 302.2, *Agricultural Engineers Yearbook,* American Society of Agricultural Engineers, St. Joseph, Mich., 1979.
8. Drainage Manual, 1st ed., Bureau of Reclamation, U.S. Department of Interior, Washington, D.C., 1978.
9. **Urquhart, L. C., Ed.,** *Civil Engineering Handbook,* 4th ed., McGraw-Hill, New York, 1959.
10. Hydrology, National Engineering Handbook, Section 4 (Part 1), Watershed Planning, U.S. Soil Conservation Service, Washington, D.C., 1964.
11. **Schwab, Glenn, O. et al.,** *Soil and Water Conservation Engineering,* John Wiley & Sons, New York, 1966.
12. Design of Open Channels, TR-25, U.S. Soil Conservation Service, Washington, D.C., 1977.
13. Ohio Erosion Control and Sediment Pollution Abatement Guide, Bull. 594, Cooperative Extension Service, Ohio State University, Columbus, 1979.

SUBSURFACE DRAINAGE SYSTEMS

W. D. Lembke and F. C. Ives

INTRODUCTION

Subsurface drainage is the removal of excess gravitational (ground) water from below the ground surface. Subsurface drainage lowers both the temporary and permanent high water table caused by the downward seepage of rainfall or irrigation water; the lateral seepage of ground water from higher lands, drainage ditches or irrigation canals; and from ground water under artesian pressure. The discussion of subsurface drainage systems in this chapter will be directed toward those with some technical background who attempt to solve field drainage problems. In many areas, the general characteristics of the soil are available through experience and are documented by various public agencies. In the discussion of systems problems, such information will be assumed to be available.

Figure 1 shows the general location of arid and humid regions in the U.S. Drainage problems in arid areas usually arise out of irrigation losses and the need for salt removal, while in the humid regions excess rainfall creates the need to remove water artificially. The objective of an agricultural drainage system should be to provide an environment for plants and cultural operations that optimizes returns in balance with the cost of the system.

THE SUBSURFACE DRAINAGE PROBLEM FOR ARID IRRIGATED SOILS

In arid irrigated areas, saline or alkali soils can develop with an improper balance of soluble salts. Saline soils occur when sodium is not a problem, which is when the conductivity of the saturation extract is more than 4 mmho/cm at 25°C and the exchangeable sodium percentage is less than 15. With adequate drainage, the excessive soluble salts can be removed by leaching through the drainage system. When the exchangeable sodium percentage becomes greater than 15, structural problems occur and the soil becomes saline-alkali or alkali depending on the concentration of soluble salts. In this chapter, discussion will be limited to the salinity problem. The reader is referred to the procedures suggested in U.S. Department of Agriculture Handbook 60 and subsequent publications of the U.S. Salinity Laboratory staff when dealing with alkali problems.

Drainage Coefficient

The approach used in designing most drainage systems involves determining a constant water removal rate on an area basis, the drainage coefficient. The units used are either millimeters per day or inches per day. For irrigated arid land, the removal rate includes irrigation losses and the planned losses for leaching and removal of salt. The equation

$$D_i C_i = D_d C_d \tag{1}$$

where D_i = depth of irrigation water, C_i = electrical conductivity* of irrigation water, D_d = depth of drainage water, C_d = electrical conductivity of drainage water, and $\dfrac{C_i}{C_d}$ = leaching requirement, is used to determine the leaching losses needed in an irrigated area. Chapter 6 can be used to determine D_i. C_i is obtained from water quality information available for the irrigation supply. C_d is found by using the permissible salinity level in the root zone

---

* Electrical conductivity is measured in mmho/cm at 25°C.

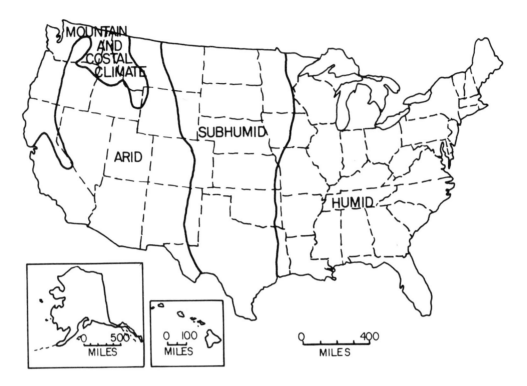

FIGURE 1.   Climatic regions of the U.S.

for some acceptable yield reduction. This salinity level is specified for a saturation extract taken from the root zone of plants. The electrical conductivity can be interchanged with salt concentrations on both sides of the above equation since these quantities are directly related. Table 1 can be used to determine the yield reduction to be expected for various crops with given levels of C_d as measured in the saturated extract of soils from research plots.

An equation for determining the drainage coefficient in an irrigated area is

$$q = \frac{\left(\dfrac{P + C}{100}\right) D_i}{F} \tag{2}$$

where q = drainage coefficient (mm/day), P = deep percolation for the leaching requirement (percent), C = irrigation losses including all those losses in the area being drained (percent), D_i = irrigation application (mm), and F = irrigation frequency. The following example illustrates determination of a drainage coefficient in an arid irrigated area.

Corn is to be irrigated in a climate where the consumptive use is 700 mm over the growing season of 4 months. The irrigation water has an electrical conductivity of 0.8 mmho/cm and the irrigation system losses other than leaching are 8%. What should be the design drainage coefficient q? A yield reduction of 10% for corn will be tolerated. The solution can be obtained with the following steps:

1. Determine C_d from Table 1. For corn with a 10% yield reduction, use C_d = 5 mmho/cm.
2. Use Equation 1 to find the leaching requirement with C_i = 0.8 mmho/cm

Table 1
SALT TOLERANCE OF PLANTS

Crops	$C_d \times 10^3$ mmho/cm at 25°C at which yield decreased by[a]		
	10%	25%	50%
Forage crops			
Bermudagrass[b] (*Cynodon dactylon* [L.] Pers.)	13	16	18
Tall wheatgrass (*Agropyron elongatum* [Host] Beauv.)	11	15	18
Crested wheatgrass (*Agropyron desertorium* [Fisches Link] Schult.)	6	11	18
Tall fescue (*Festuca arundinacea* Schreb.)	7	10.5	14.5
Barley, hay (*Hordeum vulgare* L.)	8	11	13.5
Perennial ryegrass (*Lolium perenne* L.)	8	10	13
Hardinggrass (*Phalaris stenpotera* Hack.)	8	10	13
Narrow-leaf birdsfoot trefoil (*Lotus tenuifolius* [L.] Reich)	6	8	10
Beardless wildrye (*Elymus tritticoides* Buckley)	4	7	11
Alfalfa (*Medicago sativa* L.)	3	5	8
Orchardgrass (*Dactylis glomerata* L.)	2.5	4.5	8
Meadow foxtail (*Alopecuras pratensis* L.)	2	4.5	6.5
Alsike and red clovers (*Trifolium hybridum* L. and *T. pratense* L.)	2	2.5	4
Field crops			
Barley, grain (*Hordeum vulgare* L.)	12	16	18
Sugarbeet[c] (*Beta vulgaris* L.)	10	13	16
Cotton (*Gossypium hirsutum* L.)	10	12	16
Safflower (*Carthamus tinctorius* L.)	8	11	12
Wheat (*Triticum aestivum* L.)	7	10	14
Sorghum (*Sorghum vulgare* Pers.)	6	9	12
Soybean (*Glycine max* [L.] Merr.)	5.5	7	9
Sebania (*Sebania macrocorpa* Muhl.)	4	5.5	9
Sugarcane (*Saccharum officinarum* L.)	3	5	8.5
Rice, paddy[d] (*Oryza sativa* L.)	5	6	8
Corn (*Zea mays* L.)	5	6	7
Broad bean (*Vicia faba* L.)	3.5	4.5	6.5
Flax (*Linum usitatissiumum* L.)	3	4.5	6.5
Field bean (*Phaseolus vulgaris* L.)	1.5	2	3.5
Vegetable crops			
Beet (*Beta vulgaris* L.)	8	10	12
Spinach (*Spinacia oleracea* L.)	5.5	7	8
Tomato (*Lycopersicon esculentum* Mill.)	4	6.5	8
Broccoli (*Brassica oleracea* var. *italica* L.)	4	6	8
Cabbage (*Brassica oleracea* var. *capitata* L.)	2.5	4	7
Potato (*Solanum tuberosum* L.)	2.5	4	6
Sweet corn (*Zea mays* L.)	2.5	4	6
Sweet potato (*Ipomea batatas* [L.] Lam.)	2.5	3.5	6
Lettuce (*Lactuca sativa* L.)	2	3	5
Bell pepper (*Capsicum annum* L.)	2	3	5
Onion (*Allium cepa* L.)	2	3.5	4
Carrot (*Daucus carota* L.)	1.5	2.5	4
Green bean (*Phaseolus vulgaris* L.)	1.5	2	3.5

[a]   In gypsiferous soils, EC_es causing equivalent yield reductions will be about 2 mmho/cm greater than the listed values.

[b]   Average for different varieties. Suwannee and coastal bermudagrasses are about 20% more tolerant and common and Greenfield are about 20% less tolerant than the average. For most crops varietal differences are relatively significant.

[c]   Sensitive during germination. Salinity should not exceed 3 mmho/cm.

[d]   Less tolerant during seedling stage. Salinity at this stage should not exceed 4 or 5 mmho/cm, EC_e.

From Bernstein, with permission.

$$\frac{C_i}{C_d} = \frac{0.8}{5} = 0.16 \text{ or } 16\%$$

3. Find the depth of irrigation water D_i. D_i can be found using:

$$D_i = \text{consumptive use} + \text{leaching water} + \text{losses}$$

or

$$D_i = 700 + (0.16 + 0.08) D_i$$

$$D_i = 920 \text{ mm}$$

4. Find F with an irrigation season of 4 months (assume May, June, July, and August)

$$F = 123 \text{ days}$$

5. Find q using Equation 2

$$q = \frac{(0.24)\ 920}{123} = 1.8 \text{ mm/day}$$

This solution must be modified when natural rainfall is a factor. D_i is increased by the amount of rainfall and C_i is decreased by the diluting effect. When natural rainfall is sufficient to cause water table problems, the drainage system should be designed by procedures for humid regions.

Depth and Spacing

Harmful salt concentrations in the root zone would be created by a water table as high as 24 to 30 in. below the ground surface; therefore soils in irrigated areas require subsurface drains at least 5 to 7 ft deep to control salinity.

Drainage depth and spacing criteria required for various crop-soil combinations can be obtained from local drainage guides. Drainage guides are developed from empirical data collected through evaluation of existing drainage systems. Care must be used in transposing the use of the material presented in drainage guides from one locality or soil type to another, since the empirical data are based substantially on experience and assessments of local interrelating factors.

THE SUBSURFACE DRAINAGE PROBLEM FOR HUMID AREAS

The purpose of drainage in a humid area is to lower the water table to a point below the ground surface where it will not interfere with plant root growth and development or cultural practices. The soil moisture content is substantially greater than the field capacity for a significant height above the water table; therefore plant root growth is affected to a height somewhat above the water table. Crop requirements and soil should determine the minimum depth at which the water table is maintained during the growing season. Local drainage guides are useful in determining depth and spacing, but where the water is relatively pure and there is a natural excess of water over plant requirements, the depth of drains is generally 3 to 5 ft. Table 2 shows how drain spacing is related to permeability.

The starting point in planning a subsurface drainage system is the location of the outlet. Drains may outlet by gravity or pumping into natural or artificial channels or existing tile

Table 2
TILE SPACING VS. SOIL PERMEABILITY

Permeability	Rate (in./hr)	Spacing (ft)
Slowly	0.06—0.2	30—70
Moderately slowly	0.2—0.6	60—80
Moderate	0.6—2	80—100
Moderately rapid	2—6	100—200
Rapidly	>6	200—300

Table 3
DRAINAGE COEFFICIENTS FOR HUMID REGIONS

Soil	No surface water admitted directly (mm/day)	Surface water admitted directly (mm/day)	
		Blind inlets	Open inlets
Field Crops			
Mineral	10—13	13—20	13—25
Organic	13—20	20—25	25—38
Truck Crops			
Mineral	13—20	20—25	25—38
Organic	20—38	38—51	51—102

From ASAE EP 260.3, American Society of Agricultural Engineers, St. Joseph, Mich., 1981. With permission.

outlet mains. Any of these are suitable provided they are deep enough and of sufficient capacity to remove all the drainage water from the tile line. Before proceeding with the design of the system, the adequacy of the outlet should be determined. Chapter 5 describes the design procedure for outlet drains.

Drainage Coefficients

It is common practice, when designing subsurface drains for humid areas, to use a narrow range of drainage coefficients within a broad range of climate. Table 3 is an example of a guide for selecting drainage coefficients in humid regions of the U.S. Generally, the higher values, in each range, are used in regions of higher growing season rainfall.

Determining the Size of Subsurface Pipe Drains

The size of pipe drains is determined by a required hydraulic capacity based on Manning's Equation:

$$v = \frac{1}{n} R^{2/3} S^{1/2}$$

where v = average velocity (m/sec), S = the tile gradient (dimensionless), R = the hydraulic radius ($r/2$ where v is the inside radius of the tubing) (m), and n = Manning's roughness n value. Manning's roughness n values for common materials are given in Table 4.

Table 4
MANNING *n* FOR VARIOUS MATERIALS

Type of conduit	Nominal diam. (mm)	n-value
Clay or concrete drain tile	76—760	0.013—0.014
Corrugated plastic tubing	76—203	0.015—0.016
	254—381	0.017—0.018
Corrugated metal pipe	76—760	0.025
Smooth wall pipe (tongue and groove or bell spigot joints)	76—760	0.012

Since velocity with a given pipe size determines the flow rate, Figures 2 and 3 give solutions to Manning's equation for roughness n values of 0.013 and 0.016, respectively. A sample problem is illustrated in Figure 4. This subsurface drainage system will consist of corrugated plastic tubing. Each of the lateral drains will be 700 m long with a spacing of 40 m and a grade of 0.001 m/m. The main will have a grade of 0.0015 m/m.* The required drainage coefficient is 10 mm/day. The design requires determination of the drain pipe sizes. A tabular presentation as shown in Table 5 is useful for organization of the solution.

Open Ditches for Subsurface Drainage

Open ditches may substitute for subsurface drains. The spacing and depth of such ditches follow closely the spacing and depth of pipe drains. Such ditches have the advantage of needing only a very slight grade. They are also advantageous in unstable soils where pipe drains will settle out of alignment. Open ditch systems have several disadvantages: (1) they interfere with cultural operations, (2) they require more maintenance than subsurface systems, and (3) they require considerable area that could otherwise be used for crops. In practice, open ditches are generally found only where high costs, capacity, or special conditions make pipe drains unsuitable.

SPECIAL SUBSURFACE DRAINAGE SYSTEMS

Where unusual problems are encountered, special subsurface drainage systems may be needed. Where there is no adequate outlet, a pump drainage system may be recommended. Where lateral seepage is a problem, such as where seepage water occurs on sloping land, interception drains may be needed. Mole drain systems may be used in certain heavy textured mineral soils where mole cavities will not deteriorate rapidly. Vertical drains may be used as outlets where there is a suitable aquifer for absorbing the drainage flow and where other outlets are not suitable.

Pumped-well drainage systems may be used to effectively drain land in areas with geologic conditions that are favorable. These systems have the advantage of lowering the water table to greater depths in the vicinity of the well. They may also be used as an irrigation water supply. Pumped well systems, however, have the disadvantage of requiring a higher operating cost.

---

* The spacing of the lateral lines is determined by local recommendations. The grades and lengths of the laterals and main are determined by the topography.

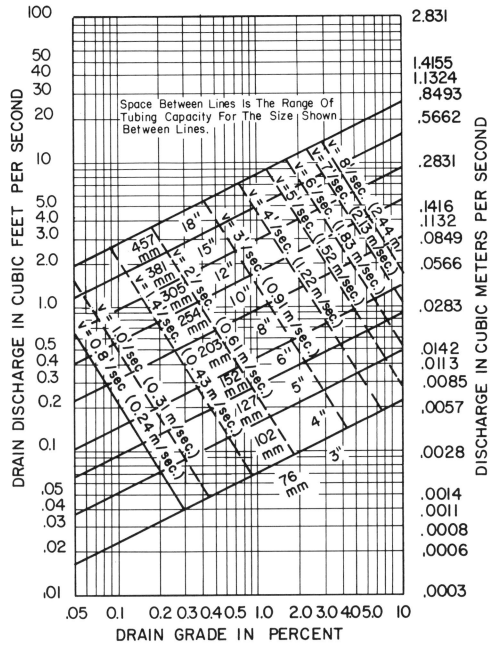

Drain Grade = Hydraulic Gradient
(Percent = Ft. Per 100 Ft, Or Meters Per 100 Meters)
V= Velocity In Feet Per Second (Meters Per Second)

FIGURE 2.   Guide for determining the required size of clay and concrete drain tile (n = 0.013).

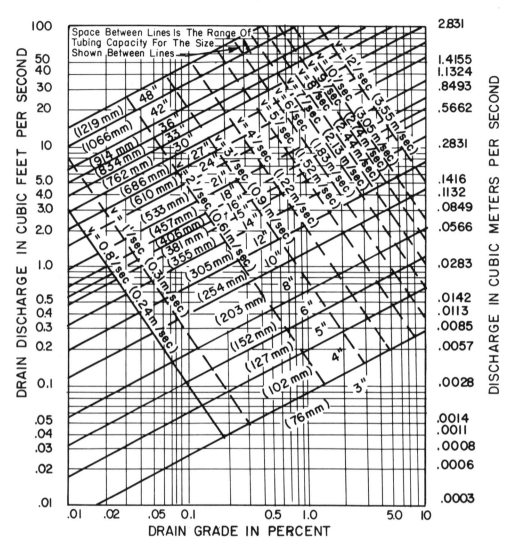

Drain Grade = Hydraulic Gradient
(Percent = Ft. Per 100 Ft, Or Meters Per 100 Meters)
V = Velocity In Feet Per Second (Meters Per Second)

FIGURE 3. Guide for determining the required size of corrugated plastic drainage tubing (n = 0.016).

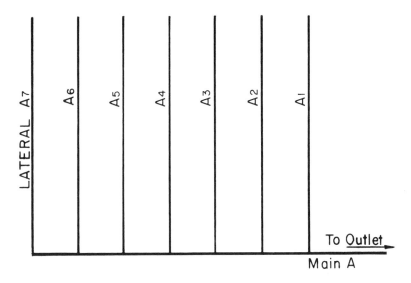

FIGURE 4.   Layout of a typical subsurface drainage system.

Table 5
SOLUTION TO SUBSURFACE DRAINAGE DESIGN PROBLEM

Line		Area drained	Required discharge	Slope	Pipe size	Length
Lat.	Main	(ha)	(m³/sec)	(m/m)	(mm)	(m)
A7		2.8	0.00324	0.001	152	700
	A	2.8		0.0015	152	40
A6		2.8			152	100
	A	5.6	0.00648		203	40
A5		2.8			152	700
	A	8.4	0.00972		203	40
A4		2.8			152	700
	A	11.2	0.0130		259	40
A3		2.8			152	700
	A	14.0	0.0162		254	40
A2		2.8			152	100
	A	16.8	0.0194		254	40
A1		2.8			152	700
	A	19.6	0.0227		305	40

DRAINAGE MATERIALS

Carroll J. W. Drablos

INTRODUCTION

During the past 20 years, technological advances associated with drainage materials have had a widespread effect on the agricultural drainage industry. Materials used for drains and accessories include clay, concrete, bituminized fiber, plastics, asbestos cement, aluminum alloy, and steel. Factors involved in selecting the type of material to be used include climatic conditions, chemical characteristics of the soil, depth requirements, and installation cost. In northern areas, freezing and thawing conditions must be considered in selecting the drain. Where acids or sulfates exist in the soil or drainage water, drains that will resist these conditions must be selected. Method of installation and depth requirements influence the selection of the type and strength of the drain. The drain must sustain the loads to which it will be subjected.

CONDUIT

A drainage conduit has three functions: to stabilize the drainage channel, to permit access of drainage water, and to carry this water efficiently to the outlet. Any conduit material selected to provide all three functions must have proper durability, adequate crushing strength, resistance to breakage, adequate water inlet area, required flow-carrying capacity, and proper installation. A drainage conduit must last for the planned life of the drainage system. The principal materials used for drainage conduits are clay, concrete, and plastic. Clay and concrete, which have been the principal drainage materials for many years, are much improved today over those used in the early 1900s. In the 1940s people became interested in plastic drain tubing, but it was not used on a commercial basis until 1967.

Material specifications for drain conduits are very important to both the drainage contractor and the landowner, since the specifications provide a basis for manufacturing a consistent product. They also ensure that the product will be strong and durable and will perform adequately in the system. Material specifications for clay, concrete, and plastic have been developed and published by the American Society for Testing and Materials (ASTM).[1-5] Engineering standards, developed by the Soil Conservation Service for subsurface drains, also identify a number of acceptable materials for obtaining durable installations (Table 1).[6]

Clay Drain Tile

The accomplishments of the clay tile industry have been outstanding in product development. Quality control has resulted in excellent products having uniformity in diameter and roundness and in length; sharp, clean-cut edges; improved density; low absorption; and uniform bearing strengths.[7,8] The two primary factors determining the durability of rigid drain tile are strength and density. The tile must be strong enough to support loads imposed by the soil and heavy machinery. The density must be high so that the tile will absorb little or no water. Dense-walled tile, because of its low absorption, resists deterioration and has a comparatively longer life than tile having high absorption. Water flows into the tile line through joints that separate the individual tiles.

ASTM Standard Specification C-4, which is currently applicable to clay drain tile,[1] separates the physical test requirements into three classes: standard, extra-quality, and heavy-duty. Standard drain tile is satisfactory where the tiles are laid in trenches with moderate

Table 1
MATERIALS FOR SUBSURFACE DRAINAGE

Type of material	Specification[a]
Clay drain tile	ASTM[b]-C-4
Clay drain tile, perforated	ASTM-C-498
Clay pipe, perforated, standard, and extra-strength	ASTM-C-700
Clay pipe, testing	ASTM-C-301
Concrete drain tile	ASTM-C-412
Concrete pipe for irrigation or drainage	ASTM-C-118
Concrete pipe or tile, determining physical properties of	ASTM-C-497
Concrete sewer, storm drain, culvert pipe	ASTM-C-14
Reinforced concrete culvert, storm, drain, sewer pipe	ASTM-C-76
Perforated concrete pipe	ASTM-C-444
Portland cement	ASTM-C-150
Asbestos-cement storm drain pipe	ASTM-C-663
Asbestos-cement nonpressure sewer pipe	ASTM-C-428
Asbestos-cement perforated underdrain pipe	ASTM-C-508
Asbestos-cement pipe, testing	ASTM-C-500
Pipe, bituminized fiber (and fittings)	Federal Spec.
Homogeneous perforated bituminized fiber pipe for general drainage	ASTM-D-2311
Homogeneous bituminized fiber pipe, testing	ASTM-D-2314
Laminated-wall bituminized fiber perforated pipe for agricultural land and general drainage	ASTM-D-2417
Laminated-wall bituminized fiber pipe, physical testing of	ASTM-D-2315
Styrene rubber plastic drain pipe and fittings	ASTM-D-2852
PVC sewer pipe and fittings	ASTM-D-2729
PVC pipe	ASTM-D-3033 or D-3034
Corrugated polyvinyl chloride tubing and compatible fittings	ASTM-F-800
Corrugated polyethylene tubing and fittings	ASTM-F-405
Corrugated polyethylene tubing and fittings 10—15 in.	ASTM-F-667
Pipe, corrugated (aluminum alloy)	Federal Spec. WW-P-402
Pipe, corrugated (iron or steel, zinc coated)	Federal Spec. WW-P-405

[a] Subsurface Drain, Standard and Specifications, Technical Guide Section IV (606), Soil Conservation Service, U.S. Department of Agriculture, Champaign, Ill., October 1980.

[b] American Society for Testing and Materials, 1916 Race Street, Philadelphia, Pa. 19103.

depths and widths and where exposure conditions are not severe. Extra-quality and heavy-duty tile should be used for more severe conditions.

To be rated as standard drain tile, clay drain tile 4 to 12 in. in diameter must have a crushing strength of 800 lb or more per foot of length (Table 2). These tiles must have an absorption that does not exceed 13% for an average of five tiles. To be rated as extra-quality, clay drain tile 4 to 14 in. in diameter must support at least 1100 lb/ft by the three-edge-bearing test and have an absorption rate of not more than 11%. Heavy-duty tile, with the same requirements for absorption rate as the extra-quality tile, can support greater loads.

Tile manufactured from shale usually has lower absorption rates and is more resistant to freezing and thawing than tile from surface clays. The quality of clay tile is also dependent upon the manufacturing process. Many plants use de-airing equipment, which increases the density of the extruded material. Generally, color and salt glazing are not reliable indicators of tile quality. Clay tiles are not affected by acid or sulfates. Low temperatures normally will not affect the use of clay tile, provided that it is properly selected for absorption and that care is taken in handling and storing the tile during freezing weather.[9]

Table 2
ASTM PHYSICAL TEST REQUIREMENTS FOR CLAY DRAIN TILE

Internal diameter (in.)	Standard drain tile				Extra-quality tile				Heavy-duty drain tile			
	Minimum crushing strength (lb/lin ft)		Maximum water absorption by 5-hr boiling (%)		Minimum crushing strength (lb/lin ft)		Maximum water absorption by 5-hr boiling (%)		Minimum crushing strength (lb/lin ft)		Maximum water absorpiton by 5-hr boiling (%)	
	Avg of 5	Indiv	Avg of 5	Indiv	Avg of 5	Indiv	Avg of 5	Indiv	Avg of 5	Indiv	Avg of 5	Indiv
4	800	680	13	16	1100	990	11	13	1400	1260	11	13
5	800	680	13	16	1100	990	11	13	1400	1260	11	13
6	800	680	13	16	1100	990	11	13	1400	1260	11	13
8	800	680	13	16	1100	990	11	13	1500	1350	11	13
10	800	680	13	16	1100	990	11	13	1550	1400	11	13
12	800	680	13	16	1100	990	11	13	1700	1530	11	13
14	840	720	13	16	1100	990	11	13	1850	1660	11	13
15	870	740	13	16	1150	1030	11	13	1980	1780	11	13
16					1200	1080	11	13	2100	1890	11	13

Concrete Drain Tile

Concrete drain tile of good quality will give long and satisfactory service under most field conditions. To ensure quality, the ASTM has established a standard.[2] There are four classes of concrete drain tile (Table 3):

1.   Standard-quality — Intended for land drainage of ordinary soils where the tiles are laid in trenches of moderate depths and widths. Tiles of this quality are not recommended where internal diameters in excess of 12 in. are required.
2.   Extra-quality — Intended for land drainage of ordinary soils where the tiles are laid in trenches of considerable depths or widths or both.
3.   Heavy-duty extra-quality — Intended for land drainage of ordinary soils where the tiles are laid in trenches of relatively great depths or widths or both.
4.   Special-quality — Intended for land drainage where special precautions are necessary for concrete tile, for example where laid in soils that are markedly acid or that contain unusual quantities of sulfates and where laid in trenches of considerable depths or widths or both.

Physical test requirements developed by ASTM for each of the classes can be found in ASTM Standard Specification C-412.[2]

Chemically, Portland cement is a base. Certain constituents of this cement in drain tile will react with the acids present in some soils. The extent of the action will depend principally on the degree of soil acidity and the permeability of the tile walls. A soil with a pH value that approaches 6.0 on the pH scale may be considered definitely acid. Concrete tile installed in definitely acid soils should be of the special-quality classification.

When concrete tiles are exposed to soils that carry an excess of about 3000 ppm (0.30%) of sodium or magnesium sulfate singly or in combination, special manufacturing precautions should be taken. In addition to specifying a special-quality tile, the cement should be of a sulfate-resistant type.[10]

When ASTM Specification C-412 is complied with, the chances are negligible for tile failures due to freezing and thawing, the action of acid and sulfate on the concrete, or soil pressures in deep and wide trenches.

Table 3
ASTM PHYSICAL TEST REQUIREMENTS FOR CONCRETE DRAIN TILE

Internal diameter (in.)	Standard-quality				Extra-quality				Heavy-duty extra-quality				Special-quality		
	Three-edge bearing crushing strength[a] (lb/lin ft) Minimum		Absorption boiled 5 hr (%) Maximum		Three-edge bearing crushing strength[b] (lb/lin ft) Minimum		Absorption boiled 5 hr (%) Maximum		Three-Edge bearing crushing strength[b] (lb/lin ft) Minimum		Absorption boiled 5 hr (%) Maximum		Three-edge bearing crushing strength[b] (lb/lin ft) Minimum	Absorption boiled 5 hr (%) Maximum	
	Avg	Indiv	Avg	Indiv	Avg	Indiv	Avg	Indiv	Avg	Indiv	Avg	Indiv	Indiv	Avg	Indiv
4	800	700	10	11	1100	990	9	10	1300	1170	9	10	1100	8	9
6	800	700	10	11	1100	990	9	10	1300	1170	9	10	1100	8	9
8	800	700	10	11	1100	990	9	10	1300	1170	9	10	1100	8	9
10	800	700	10	11	1100	990	9	10	1400	1260	9	10	1100	8	9
12[c]	800	700	10	11	1100	990	9	10	1500	1350	9	10	1100	8	9
14					1100	990	9	10	1750	1580	9	10	1100	8	9
16					1100	990	9	10	2000	1800	9	10	1100	8	9
18					1200	1080	9	10	2250	2030	9	10	1200	8	9
20					1330	1200	9	10	2500	2250	9	10	1330	8	9
22					1460	1320	9	10	2750	2470	9	10	1460	8	9
24[d]					1600	1440	9	10	3000	2700	9	10	1600	8	9
36					2400	2160	9	10	4500	4050	9	10	2400	8	9

a   Drain tile meeting the above strength requirements are not necessarily safe against cracking in deep and wide trenches.

b   For crushing strengths greater than or equal to those shown in the above table, tile may be supplied using designs with increased wall thickness, higher strength concrete, or reinforcing.

c   Tile with diameters greater than 12 in. should meet the requirements specified for extra-quality, heavy-duty extra-quality, or for special-quality concrete drain tile.

d   Tile sizes of 5 in., 15 in., and 26 through 34 in. are not included in this table.

Table 4
ASTM PHYSICAL TEST
REQUIREMENTS FOR
CORRUGATED PLASTIC TUBING

Physical property specified	Standard tubing	Heavy-duty tubing
Pipe stiffness at 5% deflection (psi)	24	30
Pipe stiffness at 10% deflection (psi)	19	25
Elongation, max. %	10	5

ASTM Specification C-118, "Concrete Pipe for Irrigation or Drainage", is for pipe used for conveying irrigation water under low hydrostatic heads and for drainage. Specification C-118 is commonly used for concrete drains as installed in the western areas of the U.S. The principal requirements of these specifications are for crushing strengths.[11]

Plastic Drain Tubing

Although plastic materials for drainage have been used in the U.S. for more than 25 years,[12,13] they were not readily accepted until the development of flexible corrugated plastic tubing in the early 1960s.[14] Installation on a large scale began about 1967. Since that time the use of plastic tubing has increased rapidly. Today this tubing accounts for a large percentage of the material installed for agricultural subsurface drainage. A major reason for the rapid acceptance is the material handling characteristics. From the installer's viewpoint the main advantage is that plastic tubing is lightweight and flexible. Because corrugated plastic tubing is a flexible conduit, it gains part of its vertical soil load-carrying capacity by lateral support of the surrounding soil. This lateral support from passive resistance of the soil occurs as the drain tube deflects outward against the soil at the sides of the tube. The amount of deflection depends on the combined strength of the flexible tube, bearing strength of the surrounding soil, period of time the tubing has been installed, methods of blinding and backfilling, stretch of the tubing during installation, and trench-bottom groove angle. A small amount of deflection is acceptable to obtain the lateral support of the soil. However, excessive deflection can lead to failure. The amount of deflection that takes place is therefore an indication of the durability of the tubing.[15,16]

High-density polyethylene has been used most extensively as a base material for subsurface agricultural drains in the U.S., whereas polyvinyl chloride (PVC) is more common in Europe. ASTM Standard F-405,[3] which is the material specification for corrugated polyethylene tubing ranging from 3 to 6 in., was developed in 1974 (see Table 4). More recently, ASTM Standard F-667[4] was developed to include sizes 8 to 24 in. in diameter. A recommended practice standard was developed for subsurface installation of corrugated thermoplastic tubing for agricultural drainage or water table control.[17] A standard has also been developed for corrugated PVC tubing and fittings intended for underground applications where soil support is given to its flexible walls. Its major uses are in agricultural and other soil drainage and septic tank effluent beds.[18]

Extensive field evaluations of plastic drain tubing have been made since 1971.[15,19-21] The results have provided the basis for many of the recommendations set forth in today's standards. Failure for corrugated plastic tubing is generally characterized by excessive deflection — usually 20 to 30% — resulting in collapse of the pipe wall.[22]

According to field studies by Drablos and Schwab, the deflection measurements of 4-in. tubing at 32 sites installed for periods averaging approximately 2 years varied from 7.3 to 50.6%, with an average of 16.4%.[19] Drablos et al.[15] in a later study showed that the average deflection for 5-, 6-, and 8-in. tubing installed for a period averaging approximately 4 years was 16.3, 15.7, and 10.3%, respectively.

Watkins and Reeve[23] stated that for practical purposes the amount of deflection of the corrugated polyethylene tubing under load is equal to the strain in the sidefill material. In other words, the vertical compression in the sidefill material governs the deflection of the tubing. Therefore the selection and placement of the soil envelope is the means by which the performance of the combined tubing-envelope structure is controlled. The denser the envelope sidefill material, the less the compressibility, and hence the smaller the tubing deflection. Watkins and Reeve further noted that, from a structural point of view, corrugated polyethylene tubing can be used successfully as buried tubing for land drainage, culverts, and other services.

ENVELOPE FILTER MATERIAL

Envelope material is installed around the tubing and tile to ensure proper bedding support and to improve the flow of groundwater into the drain. In humid-type drainage, envelope material is installed for bedding support when forming a groove in the trench bottom is not feasible. In arid-type drainage, an envelope should be installed to facilitate groundwater flow into the drain. The material also serves as proper bedding support.

A filter material on the other hand is used to restrict fine particles of silt and sand from entering the drain. A sand and gravel envelope designed as a filter may be used. Artificial prefabricated filter material can also be used. Gravel envelopes, which are not normally designed as filters, do act as partial filters because they consist of well-graded material. Prefabricated nonbiodegradable filter materials such as fiberglass, spun-bonded nylon fabric, and plastic filter cloth may be used instead of a sand-and-gravel-type filter.[17]

The terms "envelope" and "filter" are often used interchangeably. Drain filters are most often needed in silts and fine sandy soils. Soils with approximately 50% or more of fine sand or coarse silt can present sedimentation problems.

A drain filter has two major functions: to prevent soil from entering the drain except for very fine particles suspended in the drainage water and to allow unrestricted entry of water. A drain filter is not successful unless both of these functions are satisfied. It is essential that the fine particles be capable of passing through the filter material and that the larger particles be retained. If fine particles are unable to pass through, soil accumulates on the filter and within its voids, causing the flow to be drastically reduced. Continued filtration results in progressive reduction in flow and eventual total blockage. Therefore the ideal filter provides just the right sized opening so that, for a particular soil, only the fine colloidal clay particles near the drain move through the filter and are discharged with the drain water. When the filter passes fine sediment, the drain should be designed with a velocity that will carry the sediment through the drain.

As fine colloidal clay and silt particles are removed, the soil permeability in the immediate vicinity of the drain is increased, thus enhancing water entry. The larger particles and soil aggregates "bridge" across the openings of the envelope and thus stabilize and restrain the soil. Water continues to enter unrestricted through the voids. An ideal envelope is designed to restrain the soil but not to become plugged.[24]

REFERENCES

1. Standard Specification for Clay Drain Tile, ASTM Designation: C-4-62, American Society for Testing and Materials, Philadelphia, 1962 (reapproved 1981).
2. Standard Specification for Concrete Drain Tile, ASTM Designation: C-412-83, American Society for Testing and Materials, Philadelphia, 1983.
3. Standard Specification for Corrugated Polyethylene (PE) Tubing and Fittings, ASTM Designation: F-405-85, American Society for Testing and Materials, Philadelphia, 1985.
4. Standard Specification for Large Diameter Corrugated Polyethylene Tubing and Fittings, ASTM Designation: F-667-84a, American Society for Testing and Materials, Philadelphia, 1984.
5. Standard Specification for Polyethylene (PE) Corrugated Pipe With A Smooth Interior and Fittings, ASTM Designation: F-892-84, American Society for Testing and Materials, Philadelphia, 1984.
6. Subsurface Drain, Standards and Specifications, Technical Guide Section IV (606), Soil Conservation Service, U.S. Department of Agriculture, Champaign, Ill., October 1980.
7. **Fouss, J. L.,** Drain tube materials and installation, Drainage for Agriculture, Agronomy 17, American Society of Agronomy, Madison, Wis., 1974, chap. 8.
8. **Edminster, T. W.,** The challenge of new technology in drainage, Proc. ASAE Conf., Drainage for Efficient Crop Production, American Society of Agricultural Engineers, St. Joseph, Mich., 1965, 43.
9. Drainage of Agricultural Land, National Engineering Handbook, Section 16, Soil Conservation Service, U.S. Department of Agriculture, May 1971, 4.
10. **Manson, P. W.,** Concrete pipe drains, Proc. ASAE Conf., Drainage for Efficient Crop Production, American Society of Agricultural Engineers, St. Joseph, Mich., December 1965, 49—50.
11. Standard Specifications for Concrete Pipe for Irrigation or Drainage, ASTM Designation: C-118-83, American Society for Testing and Materials, Philadelphia, 1983.
12. **Schwab, G. O.,** Plastic tubing for subsurface drainage, *Agric. Eng.,* 36(2), 86, 1955.
13. **Fouss, J. L. and Donnan, W. W.,** Plastic lined mole drains, *Agric. Eng.,* 43(9), 512, 1962.
14. **Fouss, J. L.,** Plastic drains and their installation, Proc. ASAE Conf., Drainage for Efficient Crop Production, St. Joseph, Mich., December 1965, 55.
15. **Drablos, C. J. W., Walker, P. N., and Scarborough, J. L.,** Field evaluation of corrugated plastic drain tubing, 3rd Natl. Drainage Symp. Proc., ASAE Publication 1-77, American Society of Agricultural Engineers, St. Joseph, Mich., December 1976.
16. **Mitchell, J. K. and Drablos, C. J. W.,** Statistical analysis of four-inch corrugated plastic tubing deflection, *Trans. ASAE,* 17(4), 685, 1974.
17. Standard Recommended Practice for Subsurface Installation of Corrugated Thermoplastic Tubing for Agricultural Drainage or Water Table Control, ASTM Designation: F-449-76, American Society for Testing and Materials, Philadelphia, 1976.
18. Standard Specification for Corrugated Poly (Vinyl Chloride) Tubing and Compatible Fittings, ASTM Designation: F-800-84, American Society for Testing and Materials, Philadelphia, 1984.
19. **Drablos, C. J. W. and Schwab, G. O.,** Field and laboratory evaluation of 4-inch corrugated plastic drain tubing, Natl. Drainage Symp. Proc., American Society of Agricultural Engineers, St. Joseph, Mich., December 1971, 20.
20. **McCandless, D. E., Jr.,** Field evaluation of 102 mm (4-inch) corrugated polyethylene tubing, *Trans. ASAE,* 19(3), 514, 1976.
21. **Schwab, G. O. and Drablos, C. J. W.,** Deflection-stiffness characteristics of corrugated plastic tubing, *Trans. ASAE,* 20(1), 1058, 1977.
22. **Fenemor, A. D., Bevier, B. R., and Schwab, G. O.,** Prediction of deflection for corrugated plastic tubing, *Trans. ASAE,* 22(6), 1338, 1979.
23. **Watkins, R. K. and Reeve, R. C.,** Structural performance of buried corrugated polyethylene tubing, presented at 30th Annu. Highway Geology Symp., Portland, Ore., August 8, 1979.
24. **Reeve, R. C.,** Synthetics at work, drainage contractor, *Agric. Book Mag.,* 4(1), 80, 1978.

CONDUIT LOADING

G. O. Schwab

INTRODUCTION

Loads on underground conduits include those caused by the weight of the soil and by concentrated loads due to the passage of equipment or vehicles. At shallow depths, concentrated loads largely determine the strength requirements of conduits; at greater depths, the load due to the soil is the more significant. Loads for both conditions should be determined, particularly where the depth of the controlling load is not known. For drain pipe, the weight of the soil usually determines the load.

TYPES OF LOADING

For purposes of analyzing loads, the two types of underground rigid conduits are ditch conduits and projecting conduits, as shown in Figure 1. Ditch conduit conditions apply to narrow trenches, and projecting conduit conditions occur in trenches wider than about two or three times the outside diameter of the tile. However, the type of loading can be determined from Equations 1 and 2. Projecting conditions generally exist when the settlement of soil prisms A and C shown in Figure 1b is greater than prism B. Since this condition applies to conduits placed under an embankment on undisturbed soil as in road fills or pond dams, the conduit projects above a relatively solid surface; hence its name.

LOAD FACTORS FOR BEDDING

The load factor is the ratio of the strength of a rigid conduit under given bedding conditions to its strength as determined by the three-edge bearing test. As shown in Figure 2, four classes of bedding conditions are generally recognized.

For nonpermissible bedding conditions, shown in Figure 2a, no attempt is made to shape the foundation or to compact the soil under and around the conduit. Such a bedding has a low load factor (1.1) and is not suitable for drainage work.

For ordinary bedding conditions (L.F. 1.5) the bottom of the trench must be shaped to the conduit for at least one half its width. The conduit should be surrounded with fine, granular soil extending at least 0.15 m above the top of the tile.

For first-class bedding, the conduit must be placed in fine, granular material for 0.6 the conduit width and should be entirely surrounded with this material extending 0.3 m or more above the top. In addition, the blinding material should be placed by hand and thoroughly tamped in thin layers on the sides and above the conduit. Although the load factor for first-class bedding is 1.9, it is seldom necessary in tile drainage work.

The concrete cradle bearing is constructed by placing the lower part of the conduit in plain or reinforced concrete. Such construction is not practical in conservation work except for earth dams or high embankments where excessive loads are encountered. This type of bedding provides the best conditions, with load factors varying from 2.2 to 3.4.

Based on a study of trench bottoms made by several ditching machines, load factors were found to vary from 0.9 to 2.5. With a keel on the bottom of a curved trencher shoe or a V-shaped bottom, a load factor of 1.5 or greater may be expected.

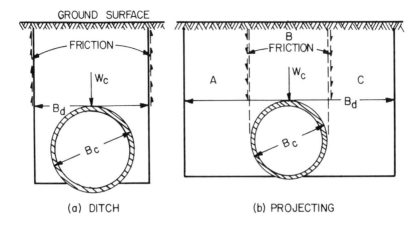

FIGURE 1.   Ditch- and projecting-type loading for rigid conduits.

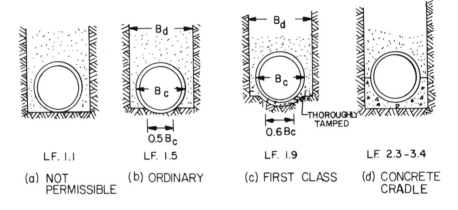

FIGURE 2.   Bedding factors for rigid ditch conduits. (From Spangler, M. G., *Am. Soc. Civil Eng. Proc.*, 73, 855, 1947. With permission.)

DESIGN FOR RIGID PIPE

Soil Loads

Loads on both ditch conduits and projecting conduits are based on studies by Marston,[1] Spangler,[2] and Schlick[3] over a period of about 40 years. For loads on ditch conduits, it is assumed that the density of the fill material is less than that of the original soil. As settlement takes place in the backfill, the sides of the ditch resist such movement (Figure 1a). Because of the upward frictional forces acting on the fill material, the load on the conduit is less than the weight of the soil directly above it. In its simplest form, the ditch conduit load formula is

$$W_c = C_d w B_d^2 \tag{1}$$

where W_c = total load on the conduit per unit length (FL^{-1}), C_d = load coefficient for ditch conduits, w = unit weight of fill material (FL^{-3}), and B_d = width of ditch at top of conduit (L).

The backfill directly over a projecting conduit (prism B in Figure 1b) will settle less than the soil to the sides of the conduit (prisms A and C). For projecting conditions, the load on

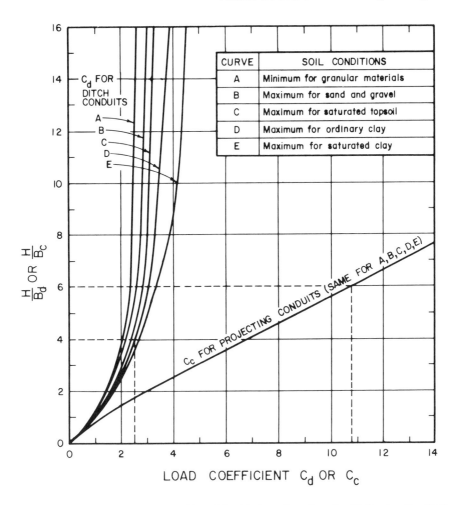

FIGURE 3.   Load coefficient for rigid conduits. (Redrawn from Spangler, M. G., *Am. Soc. Civil Eng. Proc.*, 73, 855, 1947. With permission.)

the conduit is greater than the weight of the soil directly above it, because shearing forces due to greater settlement of soil on both sides are downward rather than upward. The projecting conduit formula for wide ditches is

$$W_c = C_c w B_c^2 \tag{2}$$

where C_c = load coefficient for projecting conduits, and B_c = outside diameter of the conduit (L). The load coefficients C_d and C_c are functions of the frictional coefficient of the soil, the height of fill, and, respectively, the width of the ditch or the width of the conduit. Since these coefficients are rather complex, C_d and C_c curves, shown in Figure 3, have been developed. The load on the conduit is the smaller value as computed from Equation 1 or 2. The width of trench for a given conduit that results in the same load when computed by both equations is known as the transition width (see Figure 4).

Rigid drain pipe should be installed so that the load does not exceed the required average minimum crushing strength of the conduit. Whenever practicable, ditch conduit conditions should be provided rather than projecting conduit conditions. To prevent overloading in wide trenches, a narrow subditch can be excavated in the bottom of the main trench. The width of the subditch measured at the top of the conduit determines the load, regardless of

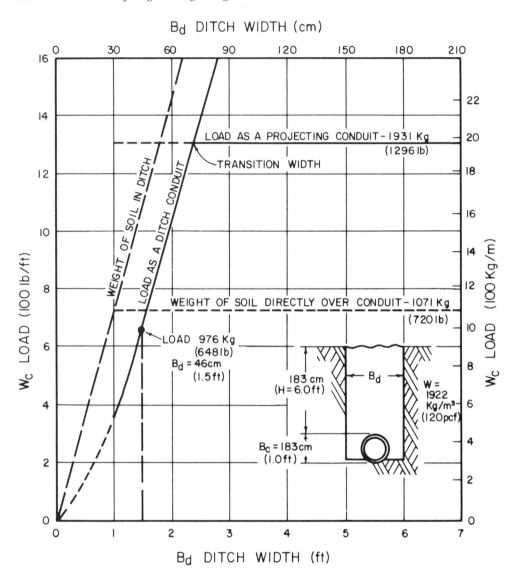

FIGURE 4.   Ditch and projecting conduit loads vs. trench width for rigid conduits. (From Schwab, G. O., Frevert, R. K., Edminster, T. W., and Barnes, K. K., *Soil and Water Conservation Engineering*, 3rd ed., John Wiley & Sons, New York, 1981. With permission.)

the shape of the trench above this point. The subditch need not extend above the top of the conduit. A common method of construction in deep cuts is to remove the excess soil with a bulldozer and to dig the subditch with a trenching machine. Allowable depths for concrete and clay pipe based on Marston's theory are given in Table 1.

Wheel Loads

Drain pipes are subjected to wheel and impact loads, especially when large tractor wheels are driven in plow furrows and heavy fertilizer or grain trucks cross fields. The percent of the live concentrated loads and impact loads transmitted to drain pipe is shown in Table 2. This percent is also called the load coefficient. Other studies by Spangler et al.[4] showed that this coefficient increases with an increase in pipe diameter.

Table 1
ALLOWABLE DEPTHS FOR RIGID PIPE IN METERS

Pipe size mm	in.	ASTM class[a]	Width of trench at top of pipe (m)						
			0.4	0.45	0.5	0.6	0.7	0.8	0.9
102—127	4—5	Std.	b	2.4	2.2	2.2	2.2	2.2	2.2
		EQ		b	3.2	3.0	3.0	3.0	3.0
152	6	Std.	b	2.7	1.9	1.9	1.9	1.9	1.9
		EQ		b	3.2	2.5	2.5	2.5	2.5
203	8	Std.	b	2.8	2.0	1.6	1.6	1.6	1.6
		EQ		b	3.3	2.4	2.2	2.2	2.2
254	10	Std.	b	2.9	2.0	1.6	1.5	1.5	1.5
		EQ		b	3.4	2.4	2.0	1.9	1.9
305	12	Std.	b	3.0	2.1	1.7	1.5	1.4	1.4
		EQ		b	3.4	2.5	2.0	1.8	1.8
381	15	EQ	b	b	3.5	2.6	2.1	1.8	1.6
457	18	EQ	b	b	4.5	3.0	2.4	2.1	1.7

Note: (1) Crushing strength based on ASTM C4 (clay tile) and C412 (concrete tile). (2) Safety factor = 1.5. (3) Unit soil weight for wet clay soil = 1.92 g/cc (120 lb/ft²). (4) Ordinary bedding with 90° bedding angle, LF = 1.5.

[a] Std. = standard-quality; EQ = extra-quality, C4 or C412.
[b] Any depth is permissible.

Table 2
PERCENT OF WHEEL LOADS TRANSMITTED TO UNDERGROUND RIGID PIPE

Depth of backfill over top of tile	Trench width at top of pipe (m)				
	0.3 (%)	0.6 (%)	0.9 (%)	1.2 (%)	1.5 (%)
0.3 m (1 ft)	17.0	26.0	28.6	29.7	29.9
0.6 m (2 ft)	8.3	14.2	18.3	20.7	21.8
0.9 m (3 ft)	4.3	8.3	11.3	13.5	14.8
1.2 m (4 ft)	2.5	5.2	7.2	9.0	10.3
1.5 m (5 ft)	1.7	3.3	5.0	6.3	7.3
1.8 m (6 ft)	1.0	2.3	3.7	4.7	5.5

Live loads transmitted are practically negligible below 1.8 m (6 ft)

These percentages include both live load and impact transmitted to 0.30 m (1 lineal ft) of pipe.

DESIGN FOR FLEXIBLE PIPE

Unlike concrete or clay tile, corrugated plastic tubing is flexible and it deflects after installation under the weight of soil above it. Part of its ability to resist failure is derived from lateral support provided by the backfill when the pipe walls deflect outward. Failure is generally characterized by excessive deflection resulting in collapse of the pipe wall. Therefore, deflection should be predicted to ensure that tubing will perform satisfactorily under given installation conditions.

For loads on flexible pipe, Spangler and Handy[5] assumed that the soil beside the tubing had essentially the same stiffness as the tubing itself and multiplied Equation 1 for ditch conditions by B_c/B_d to obtain

$$W_c = C_d w B_c B_d \tag{3}$$

This equation is valid only when thoroughly tamped soil on the sides of the tubing has essentially the same stiffness as the tubing itself. By knowing the load on the tubing, the deflection can be computed from the modified Iowa formula[5]

$$\Delta x = \frac{D_l K W_c}{(El/r^3) + 0.061 \, E'} \tag{4}$$

where Δx = horizontal deflection ($\Delta x = 0.91 \, \Delta y$) (L), D_l = deflection lag factor (usually 1.5), K = bedding angle factor (0.096 for 90° angle), W_c = soil load per unit length (FL^{-1}), r = mean radius of tubing (L), E = modulus of elasticity of tubing (FL^{-2}), I = moment of inertia of tubing (L^4), and E' = modulus of soil reaction (FL^{-2}). The two additive terms in the denominator represent the contribution of pipe stiffness and soil side support, respectively. The tubing stiffness term, El/r^3, can be more easily computed from the parallel plate stiffness test, in which

$$El/r^3 = 0.149 \, C(F/\Delta y) \tag{5}$$

where $F/\Delta y$ = parallel plate stiffness for tubing as specified in ASTM[6] F405, and C = correction factor for radius of curvature with deflection.

Field tests by Schwab and Drablos[7] on tubing varying in diameter from 102 to 381 mm showed that the load, when computed by the flexible pipe equation, was much too low, probably because soil compaction and the modulus of soil reaction did not conform to the assumptions in Equation 3. They found that tubing reaches a maximum deflection about 4 years after installation and that measured maximum deflections were about 17%. Deflections up to 20% in the field usually do not cause failure of corrugated tubing, but at 30% collapse from buckling may occur. Design should be based on 20% deflection.

Fenemor et al.[8] found that the Iowa formula using $D_l = 3.4$, $E' = 345$ kN/m², and pipe stiffness at 20% deflection gave reasonable results that agreed with field-measured deflections. They recommended maximum depths for tubing buried in loose, fine-grained soil as given in Table 3. Howard[9] developed E' values for a wider range of soils. Other design criteria may be found in ASAE[10] and Spangler and Handy.[5]

Table 3
RECOMMENDED MAXIMUM DEPTHS FOR TUBING BURIED IN LOOSE, FINE-GRAINED SOILS

Nominal tubing diam.		ASTM tubing quality	Trench width at top of tubing (m)					
mm	in.		0.2	0.3	0.4	0.6	0.8	1 or more
102	4	Standard	a	3.9	2.1	1.7	1.6	1.6
		Heavy-duty	a	a	3.0	2.1	1.9	1.9
152	6	Standard	a	3.1	2.1	1.7	1.6	1.6
		Heavy-duty	a	a	2.9	2.0	1.9	1.9
203	8	Standard	a	3.1	2.2	1.7	1.6	1.6
		Heavy-duty	a	a	3.0	2.1	1.9	1.9
254	10		a	a	2.8	2.0	1.9	1.9
305	12		a	a	2.7	2.0	1.9	1.9
381	15		a	a	a	2.1	1.9	1.9

Note: Soil modulus of reaction, $E' = 345$ kN/m²; deflection lag factor, $D = 3.4$; vertical deflection 20%; bedding angle factor, $K = 0.096$, and soil density of 1.75 g/cc.

[a] Any depth is permissible at this width or less. Minimum side clearance between tubing and trench should be about 0.08 m.

From Fenemor, A. D., Bevier, B. R., and Schwab, G. O., *Trans. ASAE,* 22(6), 1338, 1979. With permission.

REFERENCES

1. **Marston, A.,** The Theory of External Loads on Closed Conduits in the Light of the Latest Experiments, Iowa Eng. Exp. Stn. Bull. 96, 1930.
2. **Spangler, M. G.,** Underground conduits — an appraisal of modern research, *Am. Soc. Civil Eng. Proc.,* 73, 855, 1947.
3. **Schlick, W. J.,** Loads on Pipe in Wide Ditches, Iowa Eng. Exp. Stn. Bull. 108, 1932.
4. **Spangler, M. G. et al.,** Experimental Determinations of Static and Impact Loads Transmitted to Culverts, Iowa Eng. Exp. Stn. Bull. 79, 1926.
5. **Spangler, M. G. and Handy, R. L.,** *Soil Engineering,* Intext Educational Publishers, New York, 1973.
6. Standard Specifications for Corrugated Polyethylene Tubing and Fittings, ASTM F405, American Society for Testing and Materials, Philadelphia, 1976.
7. **Schwab, G. O. and Drablos, C. J. W.,** Deflection-stiffness characteristics of corrugated plastic tubing, *Trans. ASAE,* 20(6), 1058, 1977.
8. **Fenemor, A. D., Bevier, B. R., and Schwab, G. O.,** Prediction of deflection for corrugated plastic tubing, *Trans. ASAE,* 22(6), 1338, 1979.
9. **Howard, A. K.,** Modulus of soil reaction values for buried flexible pipe, *J. Geotech. Eng. Div. ASCE,* 103(GT1), 33, 1977.
10. Design and Construction of Subsurface Drains in Humid Areas, ASAE EP260.3, American Society of Agricultural Engineers, St. Joseph, Mich.,
11. **Schwab, G. O., Frevert, R. K., Edminster, T. W., and Barnes, K. K.,** *Soil and Water Conservation Engineering,* 3rd ed., John Wiley & Sons, New York, in press.

PUMP DRAINAGE

R. W. Irwin

GENERAL CONSIDERATIONS

Pump drainage is an alternative when a suitable gravity flow outlet cannot be obtained or cannot be economically developed. Pump drainage is extensively used for the drainage of agricultural land, urban areas, and sewage disposal.[1] Typical agricultural pump drainage installations are located in reclaimed areas[2,3] adjacent to oceans, lakes, bays, and rivers and in low flatland subject to periodic flooding. Many of these areas were organic soil that subsided following reclamation, leaving the general land level below the adjacent water level for a portion of the year.[4] Pumps are also used as an outlet for drainage water from tidal estuaries. Pump drainage is also used for water table control in organic soils and wildlife marshes.[5] The majority of drainage pumps are small and are used as private outlets for individual farm drainage systems in low land and for upland, such as glaciated flatland, where there is no natural outlet[6] or where the outlet would be very expensive to construct due to size or length.

PLANNING THE DRAINAGE AREA

Physical Requirements

The area to be pumped should be protected from outside waters by continuous dikes. The dikes must have a good foundation and must resist erosion from outside wave action. The dikes must also be high enough so that they are not over-topped by maximum outside water stage, and they must be wide enough so that rodents will not burrow holes through them. A collection ditch should be constructed on the inside of the dike to intercept seepage through, or under, the dike. The area to be pumped should be segregated from upland areas by means of cutoff or bypass ditches. All productive agricultural land within the pumped drained area should be uniformly tile drained to maximize economic benefits. Tile drainage systems within the area should be designed as any gravity tile drainage system. The pumping plant should maintain the water level below field tile drain outlets, except for short periods. Gravity drainage should be diverted through flood gates[7] in the dike when outside water stage permits.

Plant Location

Topography determines the site for the pumping plant. Normally, it is at the lowest elevation in the area and close to the outlet. Factors in choosing the site are availability of forebay storage, dike location, power availability, foundation condition, ease of protection from vandalism, groundwater level, and accessibility by road for construction and operation. The pumping plant should be located so that drainage water travels a minimum distance. Small drainage areas can often be served by a single pump in the lowest elevation. Larger areas may need more than one pumping plant, with separate outlets.

Collection System

The interior drainage collection net is designed in a similar manner to open drains in upland areas, except that the main collection ditches, or channels, leading to the pump should be short, straight, and direct. The channel capacity must be adequate or it must be improved so that flow of water to the pump is not restricted.

The channel hydraulic design must include sufficient velocity to provide flow at the required pumping rate. Small ditches require high velocities and need increased slope. These open drains should be deep enough to provide an outlet for interior tile drainage systems. The collection system should also provide temporary storage for the pumping plant and be easy to maintain on a regular basis.

Pumping Coefficients

A drainage coefficient is the equivalent depth of water, in millimeters, to be removed from a drainage area in 24 hr. It is used as a basis for design. A pumping coefficient is essentially the same, but can be reduced by the volume of storage available below the elevation of the land in the drainage area to be protected.

PUMPING PLANT DESIGN

Plant Capacity

Estimates should be made of the (1) average annual runoff, to estimate the annual cost of pumping and to determine the feasibility of the project; (2) seasonal distribution of runoff, to design an efficient pumping plant; and (3) maximum daily runoff, to determine the capacity of the pumping plant.

The pumping plant capacity may be determined from: (1) direct analysis using hydrologic procedures,[8,9] (2) empirical formulas[10] — these relationships should not be extrapolated beyond the area where they were developed, (3) drainage coefficients,[11,12] or (4) a study of existing installations.[13] Each method should consider the drainage and/or pumped area, volume and rate of runoff, rainfall characteristics, groundwater elevations, seepage rates, and the time of year when pumping is required. Small areas use the same normal drainage criteria as gravity drainage with a free outlet, with an allowance for available storage. The pumping plant capacity is often increased up to 20% in humid areas.[14] For large drainage areas, the plant capacity need not be as high, as all areas do not require drainage of the same degree and the value of crops often differs. The hydrologic procedure should be used for such areas.

The volume of seepage depends on the difference in elevation of the water levels outside and inside the drainage area dike. It also depends on the quality of the dike, its foundation, and subsoil. Seepage is difficult to estimate. It is usually expressed as a volume rate per unit length of dike, $m^3/sec/m$.

When runoff exceeds the capacity of the pumping plant, the excess water must be stored. Storage includes: (1) ground storage resulting from the water table rise and increase in soil water content, (2) ditch and reservoir storage, and (3) surface storage.

Drainage requirements may be expressed as a pumping coefficient or quantity of water to be removed per unit of pumped area per 24 hr.[15] The pumping coefficient is calculated from Equation 1:

$$Q_c = P - S_g - S_d - S_s + q \qquad (1)$$

where Q_c = runoff or pumping coefficient, mm/24 hr; P = 24-hr precipitation, at a selected recurrence interval, mm; S_g = ground water storage in the soil profile; mm; S_d = ditch storage, mm; S_s = storage in the forebay and sump, mm; and q = seepage in 24 hr, mm.

Pumping plant capacity may be calculated based on a 24-hr storm for the desired recurrence interval.[16] Data may be obtained from local weather stations or compiled from historic data.[17] Pumping plant capacity is the discharge or rate of flow of water, expressed in m^3/sec. The pumping plant rate may be calculated from Equation 2:

$$Q_p = \frac{0.0028 \; Q_c \; A}{T} \; m^3/sec \tag{2}$$

where Q_p = pumping plant capacity, m^3/sec; Q_c = pumping coefficient, mm/24 hr; A = area pumped, hr; and T = number of hours of pump operation per day.

Pumps — Types and Selection

Horizontal-shaft water wheels and inclined-shaft helicoid-screw pumps (Archimedes type) are used in many places in the world for low-head, high-volume discharge where excessive vegetation and trash may occur.[18] Speed of rotation is low; consequently pump wear is negligible. These pumps are not suited to freezing conditions. Headwater and tailwater elevations should also remain within narrow limits.

Centrifugal pumps are normally used for land drainage.[19] These are classified as radial-flow, mixed-flow, or axial-flow pumps. Their characteristics are shown in Figure 1. Radial-flow, or centrifugal, pumps operate at one quarter the speed of an axial-flow pump and one half the speed of a mixed-flow pump for the same head and discharge. Therefore, they can use an economical high-speed electric motor. These pumps are satisfactory for small drainage areas with high head requirements. Radial-flow centrifugal pumps are characterized by relatively low specific speeds ranging from 12 to 80.

Mixed-flow pumps are applicable for medium head operations and have medium to large capacities. They have a medium range of specific speeds from 80 to 100. Positive submergence is generally required.

Axial-flow, or propeller, pumps are best suited to land drainage as they discharge a large volume of water at low head. Axial-flow pumps have a steep characteristic between the point of maximum efficiency and zero discharge. Figure 1A shows, for a typical axial-flow pump, that at constant speed (a) as discharge increases the required power increases, (b) maximum capacity requires minimum power, and (c) zero discharge requires maximum power. These characteristics are important when choosing the driving unit for an axial-flow pump.

Figure 2 may be used to initially select the type of pump best suited to the design conditions. The actual limits of pumps are not sharply defined and areas overlap, although there is a narrow range of head for the best efficiency for each pump.

The specific speed expresses the characteristics of a pump:

$$N_s = N \; Q/H_t^{0.75} \tag{3}$$

where N_s = specific speed, rpm; N = actual pump impeller speed, rpm; Q = pump capacity, m^3/sec; and H_t = total dynamic head, m. The specific speed is the number of revolutions per minute at which a geometrically similar impeller would run if it were of such a size to deliver 1 m^3/sec against 1 m of head. The solution to Equation 3 is given in Figure 3. Impellers with low specific speed are used with pumps designed for high head operation. Pumps for low head operation usually have high specific speed impellers. The specific speed for axial-flow, or propeller, pumps ranges from 120 to 400.[20]

Net positive suction head (NPSH) is the difference between the energy level on the suction side of the pump and the vapor pressure. The pump impeller is the reference plane. Cavitation can be avoided by selection of a safe value of submergence. A high value of NPSH is required if the pump is not operating at highest efficiency. Specific speed is also a guide to the maximum suction head that will not cause cavitation.[21]

When selecting a pump, one needs to know the total dynamic head, H_t, and the pump discharge, Q, which then fixes the specific speed and type of pump to use. Choose from manufacturers' catalogs the pump that best fits the variables. Variables are maximum,

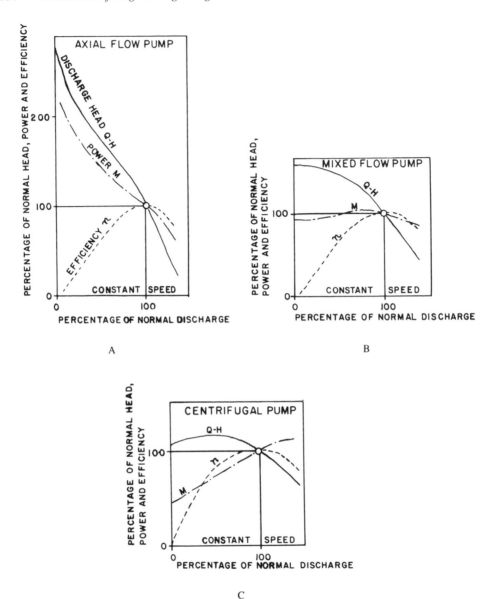

FIGURE 1.   Characteristic curves of rotary flow pumps.

minimum, and average discharge; total dynamic head based on maximum, minimum, and average stage; efficiencies; impeller diameter and speed; and type of power source to be supplied. Figure 4 shows typical characteristic performance curves for an axial-flow pump. The relationships are discharge vs. total dynamic head, power required vs. discharge, and pump efficiency vs. discharge for several impeller designations.

Water flowing through a pumping station loses part of its energy because of hydraulic losses due to friction, turbulence, and transitions. These losses are a function of velocity head, $V^2/2\,g$; therefore, flow velocity should be kept to an economical value.

In planning pumping plants, engineering surveys should be made to estimate maximum, minimum, and average static lifts. Static head, H, is the vertical distance (m) from the water level in the sump to the center line of the discharge pipe at the highest point, or the water surface of the discharge pool, whichever is greater, and is independent of losses in the

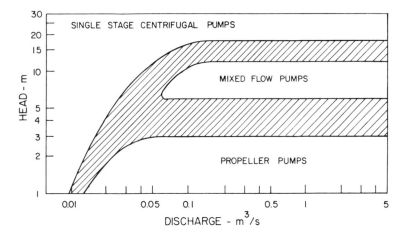

FIGURE 2. Allocation chart for drainage pumps.

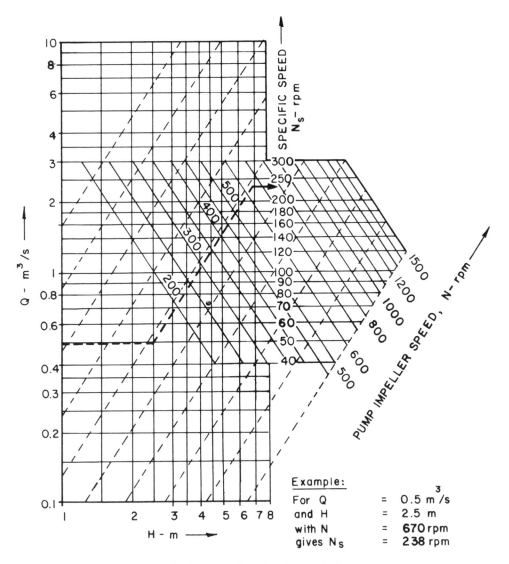

FIGURE 3. Determination of specific speed of pumps.

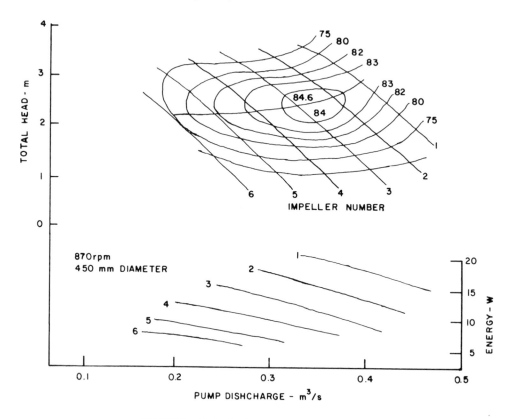

FIGURE 4.   Typical axial-flow pump relationships.

pumping station. Total dynamic head, H_t, is the static head, H, plus the velocity head, H_v, at

$$H_t = H + H_v + H_f \tag{4}$$

the discharge, plus the losses due to pipe friction, H_f, and other minor losses at flap gates, transitions, and elbows. Figures 5 and 6 can be used to estimate these losses. The losses are usually small when compared to the static head and normally do not affect the initial selection of a pump. Friction loss in discharge pipe can be reduced by designing the pipe to expand. The velocity head, H_v, may be calculated from Equation 5:

$$H_v = V^2/2 \, g = 0.083 \, Q^2/D^4 \tag{5}$$

where V = pump column velocity, m/sec; D = pump column diameter, m; Q = pump discharge, m³/sec; and g = 9.807 m/sec/sec.

The size of a pump can be estimated from Equation 6:

$$D = 1.13 \, Q^2/V^2 \tag{6}$$

where D = diameter of pump, m; Q = pump discharge, m³/sec; and V = pump column velocity, m/sec. The column velocity of an axial-flow pump may range from 1.5 to 4.0 m/sec. Use 3 m/sec as an initial estimate. Pumps operated continuously should use a lower column velocity.

The number of pumps to be installed in a pumping plant will depend upon the range of

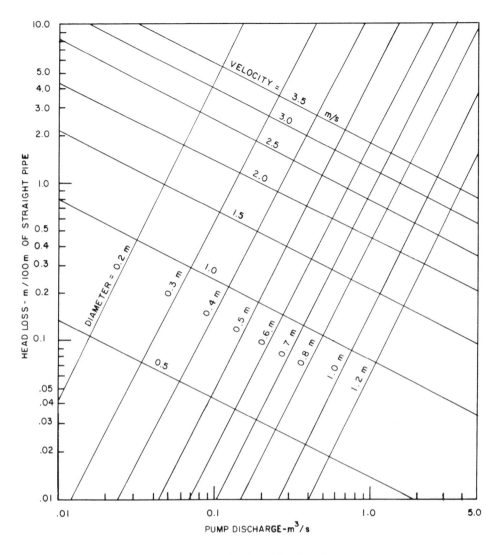

FIGURE 5. Smooth pipe friction loss chart.

flow, plant capacity, and importance of the area protected. A single pump may be used for small areas. Where the failure of a single pump may cause crop or other damage, the design flow may be divided over two or more pumps. It is common practice to install a number of pumps of the same capacity. There is also merit in using one large unit and a second unit of half the capacity for use for base flows and anticipated seepage and to conserve energy. The small pump may also be used to develop drawdown in the reservoir before the larger unit is started. In pumping plants with several pumping units, a spare unit should be considered for situations when other pumps are being maintained. The design of the pumping plant discharge capacity should have due consideration for the capacity of the outlet into which it discharges.

A pumping plant system head-capacity curve should be developed for all conditions under which the plant is to operate. The pumps are selected from manufacturers' catalogs having head-capacity characteristics that correspond as closely as possible to the overall pumping plant requirements. Based on the total dynamic head and discharge, the type of pump is selected from Figure 2. The size of the pump is determined from Equation 6. The pump is

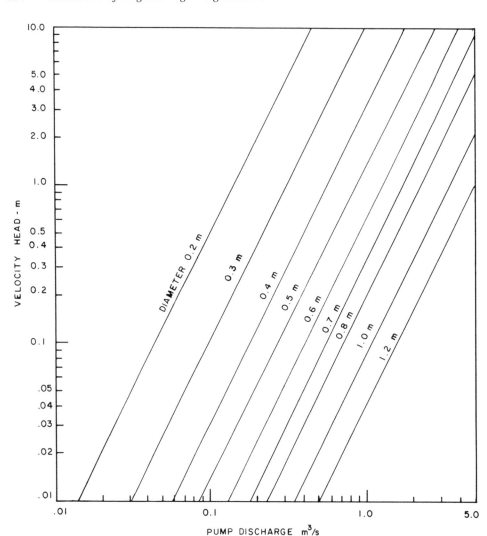

FIGURE 6.   Velocity head-discharge head relationship.

selected from manufacturers' catalogs based on capacity, total dynamic head, size, efficiency, speed, and submergence. Select pumps that will deliver the plant capacity at maximum head, which may not be the point of maximum efficiency. Select the maximum efficiency for the average operating conditions.

Storage Requirements

A stage-storage relationship should be developed for the pumped area. The storage considered in the design is that in ditches, tile drains, and low surface areas. The storage capacity of the soil should also be assessed to determine if it can be included. Each area will require individual evaluation.

Storage volume temporarily relieves the pumping plant; thus a small pump with a more economical operation can be designed.[22] Reservoirs and ditches store the dry weather flow and seepage so that the pump need not be started as frequently. Sufficient storage should be provided in the system so that a pump can be shut down temporarily for repairs, if required. It is usually less expensive to obtain more storage by increasing the horizontal

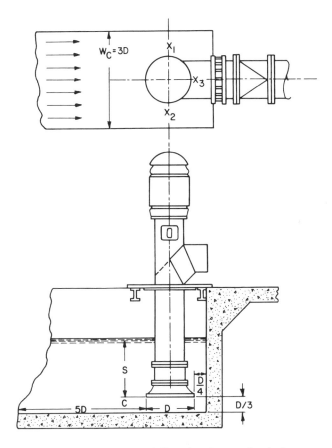

FIGURE 7. Recommended dimension clearance for single pump.

area than by increasing its depth. There should also be sufficient volume of storage so that violent changes in water level do not occur when the pump starts and stops. Storage volume may also be reduced if the site permits drainage through gravity floodgates.

Superimpose the stage-storage curve on the associated pump discharge relationship until the proper pump capacity is determined, mutually adjusted for minimum cost. A mass curve of pumping plant discharge vs. time, for an average wet year, should be developed for the drainage area.

Pumping Bays and Sumps

The forebay is the reservoir, or channel, immediately adjoining the pumping plant used for the collection and temporary storage of water. The sump is the pit, tank, or portion of the storage reservoir within the pumping plant from which the collected water is pumped.

The mean water approach velocity in the forebay and sump should be uniform across the width of the channel. The approach velocity should be low and may be estimated from:

$$V = Q/W_s(S + C) \qquad (7)$$

where the factors are illustrated in Figure 7. Channel approach velocities should not exceed 0.6 m/sec and should be reduced to 0.3 m/sec in the pumping plant. Large pumps may use higher approach velocities.[23]

Pump manufacturers' catalogs tend to illustrate rectangular sumps. Farmers tend to construct circular sumps. A circular sump is more economical to construct, but will accentuate

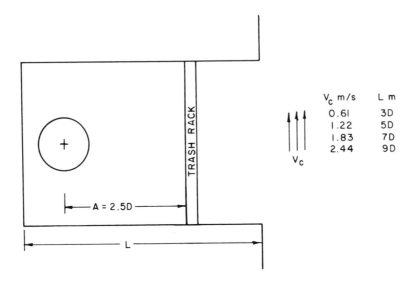

V_c m/s	L m
0.61	3D
1.22	5D
1.83	7D
2.44	9D

FIGURE 8.   Recommended dimension clearances when flow at right-angle to entrance to sump.

the rotation of water that may impair the pumping action. The submergence must be greater to suppress vortex formation.[24] Vertical baffle plates and off-center pump locations may be of some assistance in overcoming this problem. Conventional sumps are too costly for farm systems, and prefabricated sump and storage tanks have been developed.[25] These sumps may be preassembled and placed rapidly — an advantage when foundation conditions are poor.

Sump dimensions are determined by the vertical and horizontal clearances required for the pump as well as by approach conditions.[26-28] The location of the pump in a single sump should closely follow the specific dimension ratios shown in Figure 7. If D is the diameter of the bell mouth of the pump, the bottom clearance, C, should be about D/3 to D/2. The sump floor should be level for an approach distance 5D. The side clearances $X_1 = X_2 = $ D. The total width of pump bay is 3D. The spacing between the pump intake and rear wall, X_3, is about D/4. The end clearance should be smaller than the side clearance as it assists in eliminating vortex formation. The approach length, L, should be at least 4D.

The minimum clearance between the center line of adjacent pumps should be $D_1 + D_2$. (D_1 and D_2 may be of different diameters.) The minimum side clearance, X_1 and X_2, = D, and the rear wall clearance, X_3, = D/2. A bulk-head from the floor level of the sump to the top of the propeller assembly should be used between pump units to reduce flow interference.

When the flow of water in the approach channel is at right angles to the forebay, or inlet to the sump, the dimension ratios shown in Figure 8 should be used.

The minimum submergence (S) for an axial-flow pump measured from the propeller inlet to the water level is based on: (1) a minimum water level to cover the top of the propeller to keep the pump self-priming, (2) preventing cavitation — the available NPSH must be equal to or larger than the required NPSH, (3) vortex suppression — a value recommended by the manufacturer and not the same as for cavitation, and (4) providing adequate storage capacity so that the pump need not start and stop too often. All submergence should be checked against specific speed curves and manufacturers' charts. The minimum submergence, S, to avoid surface air-entraining vortices should be 1.5D.

Water level in the sump should never be allowed to drop to the extent that the minimum submergence is exceeded. For manual start operation with automatic stop, the number of recommended starts is two per day. The volume of storage, S_s m³, required for a selected pump capacity m³ can be estimated from Equation 8:[29]

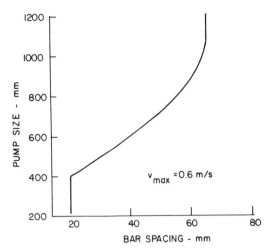

FIGURE 9.   Trash rack bar spacing for drainage pumps.

$$S_s = 11\ 235Q \tag{8}$$

For electric motor drives, there should be sufficient storage between start and stop levels to prevent excessive pump cycling. The sump and pump design should be balanced so that the minimum operating cycle would be about 6 min under critical conditions.[30]

Provision should be made for sealing off the pump sump so repairs can be made to it or the pump. Stop-logs placed in grooves in the sump wall is an economical method of doing this.

Trash racks are bar grates between the forebay and sump for excluding floating objects and debris that might damage the pump or interfere with its operation. The recommended bar sizes and bar spacings are shown in Figure 9 for each pump size. Trash racks should be at least 2.5D in front of the pump. They should have sufficient net area so that the velocity of flow through the bars does not exceed 0.6 m/sec. Trash racks should be sloped for ease of cleaning and raking. Large pumping stations should have automatic trash rack cleaners. Small pumping plants may use galvanized basket strainers. Site trash racks and screens so that they serve as flow straighteners.

Power Requirements

The required power of the driving unit, watts, is determined from:

$$W = \frac{9800\ Q\ H_t}{\eta_p \eta_g \eta_e} \tag{9}$$

where Q = pump discharge, m³/sec; H_t = total dynamic head, m; η_p = efficiency of the pump, %; η_g = efficiency of the right-angle gear drive, or other drive, %; and η_e = efficiency of the energy source.

Any energy device that produces sufficient torque on a continuous basis can be used. Power requirements should be calculated based on the least favorable conditions. Windmills are often used as an energy source for small individual pumps having a capacity of 1 to 15 ℓ/m and 1 m lift.[18]

Electric motors usually provide the most economical installation where three-phase power lines are available close to the pumping plant. Electric motors should be used as direct drive whenever possible for minimum cost and maximum efficiency. A diesel-operated pump should be used as a standby. It could also be used to reduce excessive KVA power demands.

Internal combustion engines are of two types. Spark ignition engines use gasoline, LP gas, or natural gas. Compression ignition engines use diesel fuel. Diesel engines have a higher first cost, but are preferred when the number of hours of operation justifies the additional cost. Engines should operate at 85 to 90% of full load. Engines have the advantage of variable speed to match reduced inflows and lower heads. They require more space and more maintenance.

Engines require a right-angle gear drive for larger installations where the speed of the pump and motor differ. The average efficiency of these right-angle drive units should exceed 95%. Specifications will require the direction of rotation, watts input of the pump, revolutions per minute of the pump and engine, and engine rotation and pump thrust for a specific speed ratio. Smaller pumps may utilize a belt drive or power-take-off drive. Engine drives will require space for fuel storage, batteries, cooling facilities, and control equipment.

Controls, Equipment, and Housing

Controls may include pressure gages for head level, instrumentation for flow measurement, and telemetering devices to transmit the data collected. Other controls may include automatic pump motor starters, automatic pump controls, water level controls by float (electrode or air), variable pitch axial flow pumps, and controls for pump speed variation.

Large pumping plants require automatic controls with a remote alarm signal for malfunction or abnormal conditions. Float control logic circuits automatically select the proper combination of pumps to match the discharge to the inflow. Small pumping plants use float or electrode controls for automatic operation.[15] Controls are also required for the power source.

The pumping plant must be housed in an adequate building. The foundation must support the structure evenly on each side so that the pump will hang perpendicular. The floor must support the pump base plate with the entire weight of the pump, including downthrust and vibration. The hydraulic downthrust for axial-flow pumps may be estimated from Figure 10. The discharge pipe must also be supported to avoid deflection of the pump. The floor of the house should be sloped for floor drainage, and there must also be access to the pump for repairs and hoisting equipment to provide for the handling of equipment and materials. The building must be well-ventilated and heated where required. Railings and guards must be used at dangerous spots. The site should be fenced to protect against vandalism and hazards to people.[31]

Discharge pipes may be placed over or through the dike. Those placed over the dike are less expensive to install, but result in a higher operating cost. Those placed in the dike must withstand water and earth pressure and must not create leaks due to pump vibration.

Submerged outlets for siphon action reduce the operating cost because of the lower head. Special air vents are required to break the siphon to prevent backflow when the pump stops. The siphon may be primed with a vacuum pump.

OPERATION AND MAINTENANCE

Power Consumption

The drive unit is selected based on the amount of energy required, availability and cost of energy, initial cost of the drive unit, depreciation of the unit, maintenance cost, and labor availability. When the energy requirement has been calculated from Equation 9, the energy cost can be estimated from data in Table 1. Table 1 is based on a pump efficiency of 75%.

Economics of Pump Drainage

The cost of using pump drainage for an area should be computed on an annual basis.[32,33] Factors that determine the annual fixed cost of the pumping plant include:

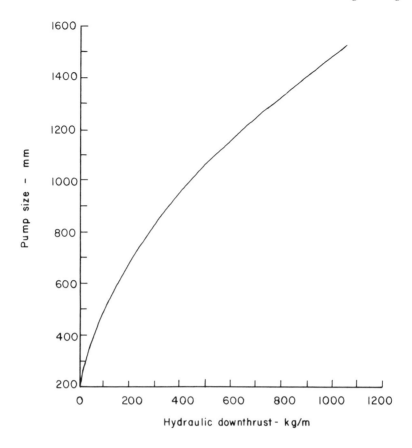

FIGURE 10.   Hydraulic downthrust for axial-flow pumps, kilograms per meter of pumping head.

Table 1
ENERGY PERFORMANCE
STANDARDS FOR
PUMPING PLANTS

Energy source	Energy cost $/kWh
Diesel	$\dfrac{\text{cost of diesel fuel/L}}{1.9}$
Gasoline	$\dfrac{\text{cost of gasoline/L}}{1.3}$
LP gas	$\dfrac{\text{cost of LP gas/L}}{1.1}$
Electric	$\dfrac{\text{cost of electricity/kWhr}}{0.9}$

1. Depreciation on the pumping plant, storage reservoir, and structure
2. Interest on the capital investment
3. Rent on the pumping plant
4. Taxes and insurance
5. Depending upon the situation, basic energy cost and labor cost and employee benefits

The annual operating cost varies with the discharge, total dynamic head, and hours of operation.[34] Factors to consider are fuel or energy, lubrication, repair and maintenance, and the labor policy. There is a curvilinear relationship between operating and maintenance costs.[34] Costs are often expressed in terms of the area drained.

REFERENCES

1. **Clay, C.,** Pumping stations, with special reference to land drainage and storm water disposal, Paper 5458, *J. Inst. Civil Eng.,* 24, 35 and 449, 1945.
2. **Woodward, S. M.,** Land Drainage by Means of Pumps, Office Exp. Stn., Bull. 243, U.S. Department of Agriculture, Washington, D.C., 1911.
3. **Irwin, R. W.,** Pump drainage outlets, *Can. Agric. Eng.,* 4, 10, 1962.
4. **Shafer, F. F.,** The Problems of Controlled Drainage for Muck and Peat Lands, Upper Mississippi Region, Soil Conservation Service, U.S. Department of Agriculture, Milwaukee, 1946.
5. **Allison, U. S.,** The practice of controlled drainage in Florida, *Assoc. South. Agric. Workers Proc.,* 55, 27, 1956.
6. **Hull, D.,** Pumps the answer to hard to drain spots, *Successful Farming,* 51(14), 150, 1953.
7. **O'Neall, W. Q.,** *Handbook of Water Control,* Chicago, 1936.
8. **Adams, H. W.,** Pumping requirements for leveed agriculture areas, *J. Irrig. Drain. Am. Soc. Civil Eng.,* 83(Ir 1), 1, 1957.
9. **Parsons, O. J.,** Pumping stations for drainage areas designed graphically, *Civil Eng.,* 23, 553, 1953.
10. **Sutton, J. G.,** Design and Operation of Drainage Pumping Plants, Tech. Bull. 1008, U.S. Department of Agriculture, Washington, D.C., 1950.
11. Drainage principles and practices, *Int. Inst. Land Reclam. Imp. Neth. Publ.,* 16(4), 199, 1974.
12. **Sutton, J. G.,** Pumping plants for land drainage, *Agric. Eng.,* 36, 243, 1955.
13. **Irwin, R. W.,** Survey of Pump Drainage Installations in Ontario, Eng. Tech. Publ. 7, Ontario Agriculture College, Canada, 1960.
14. Drainage of Agricultural Land, National Engineering Handbook, Section 16, Soil Conservation Service, U.S. Department of Agriculture, Washington, D.C., 1973.
15. ASAE, Design of agricultural drainage pumping plants, EP 369, *Agricultural Engineers' Yearbook,* American Society of Agricultural Engineers, St. Joseph, Mich., 1969, 483.
16. **Sutton, J. G.,** Design and Operation of Drainage Pumping Plants in the Upper Mississippi Valley, USDA Tech. Bull. 390, U.S. Department of Agriculture, Washington, D.C., 1933.
17. Weather Bureau, Rainfall Frequency Atlas of the United States, Tech. Paper, 40, U.S. Department of Commerce, Washington, D.C., 1961.
18. **Molenaar, A.,** Water Lifting Devices for Irrigation, UN-FAO Agric. Dev. Paper No. 60, United Nations Food and Agriculture Organization, Rome, 1956.
19. **Stepanoff, A. J.,** *Centrifugal and Axial Flow Pumps,* 2nd ed., John Wiley & Sons, New York, 1948.
20. Great Britain Ministry of Technology, Pump Design, Testing and Operation, National Engineering Lab., Her Majesty's Stationery Office, London, 1966.
21. **Iverson, H. W.,** Studies of submergence requirements of high-specific-speed pumps, *Trans. Am. Soc. Mech. Eng.,* 75, 635, 1953.
22. **Irwin, R. W.,** Ditch storage for farm drainage pumping plants, *Agric. Eng.,* 44, 548, 1963.
23. **Dicmas, J. L.,** Development of an optimum sump design for propeller and mixed flow pumps, Paper 67: FE 26, American Society of Mechanical Engineers, New York, 1967.
24. **Shahroody, A. M. and Davis, J. R.,** Efficiency of pumping from small circular sumps, *J. Irrig. Drain. Am. Soc. Civil Eng.,* 90(IR 4), 1, 1964.
25. **Sass, M. and Irwin, R. W.,** Design and Installation of Farm-Size Pumping Plants, Paper 77-2582, American Society of Agricultural Engineers, St. Joseph, Mich., 1977.
26. Standards of the Hydraulic Institute, 13th ed., Hydraulic Institute, Cleveland, 1971.
27. **Prosser, M. J.,** The Hydraulic Design of Pump Sumps and Intakes, BHRA/CIRA Report, July 1977.
28. **Fletcher, B. P. and Grace, J. L.,** Flow Conditions at Pumping Stations: Cairo, Ill., Tech. Rep. H-77-3, Hydraulics Laboratory, U.S. Army Engineer Waterways Experiment Station, Vicksburg, Miss, 1977.
29. **Larson, C. L. and Allred, E. R.,** Planning pump drainage outlets, *Agric. Eng.,* 37, 38, 1956.
30. **Larson, C. L. and Manbeck, D. M.,** Factors in drainage pumping efficiency, *Agric. Eng.,* 42, 296, 1961.
31. U.S. Army, Pumping Stations, Eng. Manual, Civil Works Construction, Part CXXXI.

32. **Irwin, R. W.,** The Cost of Land Drainage in Dover Township, Canada, Trans. 7th Int. Congr. Irrig. Drainage, Mexico City, 1969, Q 25, R 2.

33. **Duty, C. W.,** Pump drainage in Carolina Bays, *J. Irrig. Drain. Am. Soc. Civil Eng.*, 99(IR 4), 465, 1973.

34. **Eyer, J. M.,** Pumping-plant operation and maintenance costs, *J. Irrig. Drain. Am. Soc. Civil Eng.*, 91(IR 4), 37, 1965.

Waste Management

MANURE PRODUCTION AND CHARACTERISTICS*

ASAE Data 384

Data on farm animal manure production and characteristics are presented to make readily available information for the planning, design, and operation of structures and equipment for animal enterprises.

These data are combined from a wide base of published and unpublished information on animal waste production and characterization. Those making use of this information should recognize that these are median values and that actual values vary widely (by as much as 50%) due to differences in ration, animal age, and management practices.

---

* ASAE D384 — developed by the Engineering Practices Subcommittee of the ASAE Agricultural Sanitation and Waste Management Committee; approved by the Structures and Environment Division Standards Committee; adopted by ASAE December 1976; reconfirmed for 1 year December 1981, December 1982, December 1983, December 1984.

Table 1
MANURE PRODUCTION AND CHARACTERISTICS PER 454 kg (1000 lb) LIVE WEIGHT[a]

Item	Units	Dairy		Beef		Swine		Sheep	Poultry		Horse
		Cow	Heifer	Yearling 182—318 kg (400—700 lb)	Feeder >318 kg (>700 lb)	Feeder	Breeder		Layer	Broiler	
Raw waste (RW)	kg/day	37.2	38.6	40.8	27.2	29.5	22.7	18.1	24.0	32.2	20.4
	lb/day	82.0	85.0	90.0	60.0	65.0	50.0	40.0	53.0	71.0	45.0
Feces/urine ratio		2.2	1.2	1.8	2.4	1.2		1.0			4.0
Density	kg/m³	1005.0	1005.0	1010.0	1010.0	1010.0	1010.0		1050.0	1050.0	
	lb/ft³	62.7	62.7	63.0	63.0	63.0	63.0		65.6	65.6	
Total solids (TS)	kg/day	4.7	4.2	5.2	3.1	2.7	1.9	4.5	6.1	7.7	4.3
	lb/day	10.4	9.2	11.5	6.9	6.0	4.3	10.0	13.4	17.1	9.4
	% of RW	12.7	10.8	12.8	11.6	9.2	8.6	25.0	25.2	25.2	20.5
Volatile solids	kg/day	3.8			2.7	2.2	1.4	3.8	4.3	5.4	3.4
	lb/day	8.6			5.9	4.8	3.2	8.5	9.4	12.0	7.5
	% of TS	82.5			85.0	80.0	75.0	85.0	70.0	70.0	80.0
BOD₅[b]	% of TS	16.5			23.0	33.0	30.0	9.0	27.0		
COD[c]	% of TS	88.1			95.0	95.0	90.0	118.0	90.0		
TKN[d]	% of TS	3.9	3.4	3.5	4.9	7.5		4.5	5.4	6.8	2.9
P[e]	% of TS	0.7	3.9		1.6	2.5		0.66	2.1	1.5	0.49
K[f]	% of TS	2.6			3.6	4.9		3.2	2.3	2.1	1.8

[a] Numerical values for kg/day/1000 kg live weight are the same as those for lb/day/1000 lb live weight.
[b] Five-day biochemical oxygen demand.
[c] Chemical oxygen demand.
[d] Total Kjeldahl nitrogen.
[e] Phosphorus as P.
[f] Potassium as K.

DESIGN OF ANAEROBIC LAGOONS FOR ANIMAL WASTE MANAGEMENT*

ASAE EP403

1. PURPOSE AND SCOPE

1.1. This Engineering Practice describes the minimum criteria for design and construction of anaerobic animal waste lagoons located in predominantely rural or agricultural areas.

2. DEFINITION

2.1. Lagoons are impoundments made by constructing an excavated pit, dam, embankment, dike, levee, or by a combination of these procedures.

2.2. Anaerobic lagoons treat animal wastes by predominantly anaerobic biological action. Utilization of this treatment process applies where concentrated animal waste must be treated, handled, or retained to reduce sources of pollution, minimize health hazards, and maintain or improve the local environment.

3. LAWS AND REGULATIONS

3.1. Federal regulations require that animal waste management system operations result in no point source of pollutants except as the result of catastrophic or chronic rainfall events. Anaerobic animal waste lagoons are designed so that residuals of the treatment or storage process may be returned to the soil for terminal disposition. All other federal, state, and local laws, rules, and regulations governing the use of animal waste management lagoons should be followed. Necessary approvals and permits for location, design, construction, and operation of the lagoon should be secured from accountable authorities.

4. DESIGN CRITERIA

4.1. Characteristics of a successfully operating animal waste lagoon vary throughout the U.S. Odor intensity is most often the criteria of success, but rates of sludge and supernatant accumulation are also important. This Engineering Practice provides recommended design values. Due to varying opinions of acceptable characteristics, other established design values may differ from these values. These criteria recommended in this section should be compared with locally accepted design criteria. Figure 1 illustrates various construction details.

4.2. Location — For convenience the lagoon should be located adjacent to the source of waste or as near as is practical.

4.2.1. Odor — Anaerobic lagoons will produce undesirable odors at times, but these will be minimized by following recommended design and management practices. Odors are usually strongest during the changeover from winter to spring operating conditions. These odors are generally stronger in colder climates. Because of odor production, anaerobic lagoons are usually located in isolated areas. Additional information can be found in ASAE Engineering Practice EP379, Control of Manure Odors.

4.2.2. Residence — Lagoons should be located at least 90 m (300 ft) from residences or other places of occupancy or employment. This minimum separation may be extended to

---

* ASAE Engineering Practice: ASAE EP403, American Society of Agricultural Engineers, St. Joseph, Mich., 394—396. Developed by the ASAE Agricultural Sanitation and Waste Management Committee; approved by the Structures and Environment Division Standards Committee; adopted by ASAE March 1981.

ASAE Engineering Practice: ASAE EP403

DESIGN OF ANAEROBIC LAGOONS FOR ANIMAL WASTE MANAGEMENT

Developed by the ASAE Agricultural Sanitation and Waste Management Committee; approved by the
Structures and Environment Division Standards Committee; adopted by ASAE March 1981.

SECTION 1—PURPOSE AND SCOPE

1.1 This Engineering Practice describes the minimum criteria for design and construction of anaerobic animal waste lagoons located in predominantly rural or agricultural areas.

SECTION 2—DEFINITION

2.1 Lagoons are impoundments made by constructing an excavated pit, dam, embankment, dike, levee, or by a combination of these procedures.

2.2 Anaerobic lagoons treat animal wastes by predominantly anaerobic biological action. Utilization of this treatment process applies where concentrated animal waste must be treated, handled, or retained to reduce sources of pollution, minimize health hazards, and maintain or improve the local environment.

SECTION 3—LAWS AND REGULATIONS

3.1 Federal regulations require that animal waste management system operations result in no point source of pollutants except as the result of catastrophic or chronic rainfall events. Anaerobic animal waste lagoons are designed so that residuals of the treatment or storage process may be returned to the soil for terminal disposition. All other federal, state, and local laws, rules, and regulations governing the use of animal waste management lagoons should be followed. Necessary approvals and permits for location, design, construction, and operation of the lagoon should be secured from accountable authorities.

(2500 ft) in northern states. A natural or constructed screen concealing a lagoon may be desirable.

4.2.3 Wind. Lagoons should be located so that prevailing winds disperse and transport lagoon odors away from residences. Spring winds are of greatest concern.

4.2.4 Water supply. Lagoons should be located so that water supply wells are not contaminated. A minimum distance of 90 m (300 ft) from wells is recommended; applicable state or local regulations should be followed.

4.2.5 Expansion. Space should be allowed for possible expansion of the animal enterprise and the lagoon facility consistent with long range plans.

4.3 Soil and Foundation. A site investigation should be made to determine the physical characteristics and suitability of each construction site. Lagoons should be located on soils of low permeability which seal quickly. Where the wetted soil surface should be sealed, an impermeable membrane such as polyethylene film; impermeable soil from a borrow area, incorporation of bentonite clay with the soil, or other material acceptable to the responsible government agency may be used.

4.4 Size. Anaerobic lagoons for treatment of animal waste are designed on the basis of waste load added per unit volume of capacity. Operating temperatures and loading procedures are major factors determining the rate of biodegradation.

4.4.1 Volume: Fig. 2 shows recommended maximum lagoon loading rates for the United States based on average monthly temperatures and corresponding biological activity rates. The maximum loading rate is expressed in terms of grams of volatile solids

SECTION 4—DESIGN CRITERIA

4.1 Characteristics of a successfully operating animal waste lagoon vary throughout the United States. Odor intensity is most often the criteria of success, but rates of sludge and supernatant accumulation are also important. This Engineering Practice provides recommended design values. Due to varying opinions of acceptable characteristics, other established design values may differ from these values. These criteria recommended in this section should be compared with locally accepted design criteria. Fig. 1 illustrates various construction details.

4.2 Location. For convenience the lagoon should be located adjacent to the source of waste or as near as is practical.

4.2.1 Odor. Anaerobic lagoons will produce undesirable odors at times, but these will be minimized by following recommended design and management practices. Odors are usually strongest during the changeover from winter to spring operating conditions. These odors are generally stronger in colder climates. Because of odor production, anaerobic lagoons are usually located in isolated areas. Additional information can be found in ASAE Engineering Practice EP379, Control of Manure Odors.

4.2.2 Residences. Lagoons should be located at least 90 m (300 ft) from residences or other places of occupancy or employment. This minimum separation may be extended to as much as 760 m

(VS) per day per cubic meter (lb VS/day/1000 ft³) of lagoon volume. An adjusted loading rate allowing for animal species and ration is the product of the recommended maximum loading rate and the loading rate multiplier found in Table 1. Table 2 lists the daily VS production for various animal species. Related information can be found in ASAE Data D384, Manure Production and Characteristics.

4.4.1.1 Example. Problem: A central South Carolina swine producer wishes to build an anaerobic lagoon to treat the wastes from a 100-sow, farrow-to-finish, totally confined swine operation. What is the recommended volume of the lagoon?
Solution: Assume the following numbers and sizes of animals are housed:

    100 sows, ave. wt. 160 kg. (350 lb.)
      6 boars, ave. wt. 180 kg. (400 lb.)
     10 replacements, ave. wt. 115 kg. (250 lb.)
    750 pigs, ave. wt. 55 kg. (120 lb.)
    Total animal weight is 59,480 kg. (129,900 lb.)

From Table 2, volatile solids (VS) production rate is 4.8 g per day per kg. animal weight (lb/day/1000 lb.)
Total daily VS production = 4.8 × 59,480
                          = 285,500 g (624 lb)

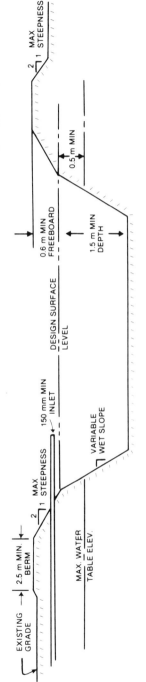

FIGURE 1. Anaerobic lagoon construction details.

FIGURE 2.   Recommended maximum loading rates for anaerobic lagoons for animal waste in g VS/day/m³ (lb/VS/day/1000 ft³).

as much as 760 m (2500 ft) in northern states. A natural or constructed screen concealing a lagoon may be desirable.

4.2.3. Wind — Lagoons should be located so that prevailing winds disperse and transport lagoon odors away from residences. Spring winds are of greatest concern.

4.2.4. Water supply — Lagoons should be located so that water supply wells are not contaminated. A minimum distance of 90 m (300 ft) from wells is recommended; applicable state or local regulations should be followed.

4.2.5. Expansion — Space should be allowed for possible expansion of the animal enterprise and the lagoon facility consistent with long-range plans.

4.3. Soil and foundation — A site investigation should be made to determine the physical characteristics and suitability of each construction site. Lagoons should be located on soils of low permeability which seal quickly. Where the wetted soil surface should be sealed, an impermeable membrane such as polyethylene film, impermeable soil from a borrow area, incorporation of bentonite clay with the soil, or other material acceptable to the responsible government agency may be used.

4.4 Size — Anaerobic lagoons for treatment of animal waste are designed on the basis of waste load added per unit volume of capacity. Operating temperatures and loading procedures are major factors determining the rate of biodegradation.

4.4.1. Volume — Figure 2 shows recommended maximum lagoon loading rates for the U.S. based on average monthly temperatures and corresponding biological activity rates. The maximum loading rate is expressed in terms of grams of volatile solids (VS) per day per cubic meter (lb VS/day/1000³) of lagoon volume. An adjusted loading rate allowing for animal species and ration is the product of the recommended maximum loading rate and the loading rate multiplier found in Table 1. Table 2 lists the daily VS production for various animal species. Related information can be found in ASAE DATA D384, Manure Production and Characteristics.

<table>
<tr><th colspan="3" align="center">Table 1
LOAD RATE MULTIPLIER FOR
ANIMAL SPECIES AND RATION</th></tr>
<tr><th>Animal
species</th><th>Ration</th><th>Loading rate
multiplier</th></tr>
<tr><td>Swine</td><td>High concentrate</td><td>1.0</td></tr>
<tr><td>Dairy</td><td>High forage</td><td>0.75</td></tr>
<tr><td>Poultry</td><td>High concentrate</td><td>0.8</td></tr>
<tr><td>Beef</td><td>High concentrate</td><td>1.0</td></tr>
<tr><td></td><td>High forage</td><td>0.75</td></tr>
<tr><td>Sheep</td><td>High concentrate</td><td>0.9</td></tr>
<tr><td>Horse</td><td>High forage</td><td>0.75</td></tr>
</table>

<table>
<tr><th colspan="2" align="center">Table 2
DAILY VOLATILE SOLIDS
PRODUCTION BY LIVESTOCK</th></tr>
<tr><th>Animal
type</th><th>Volatile solids produced, per day
per kg of animal (lb/day/1000 lb)</th></tr>
<tr><td>Swine</td><td>4.8</td></tr>
<tr><td>Dairy</td><td>8.6</td></tr>
<tr><td>Poultry</td><td></td></tr>
<tr><td>Layer</td><td>9.4</td></tr>
<tr><td>Broiler</td><td>12.0</td></tr>
<tr><td>Beef</td><td></td></tr>
<tr><td>Feeder</td><td>5.9</td></tr>
<tr><td>Sheep</td><td>8.5</td></tr>
<tr><td>Horse</td><td>7.5</td></tr>
</table>

Example — Problem: A central South Carolina swine producer wishes to build an anaerobic lagoon to treat the wastes from a 100-sow, farrow-to-finish, totally confined swine operation. What is the recommended volume of the lagoon?

Solution: Assume the following numbers and sizes of animals are housed:

- 100 sows, avg wt 160 kg (350 lb)
- 6 boars, avg wt 180 kg (400 lb)
- 10 replacements, avg wt 115 kg (250 lb)
- 750 pigs, avg wt 55 kg (120 lb)
- Total animal weight is 59,480 kg (129,900 lb)

From Table 2, VS production rate is 4.8 g/day/kg animal weight (lb/day/1000 lb). Total daily VS production = 4.8 × 59,480 = 285,000 g (624 lb). From Figure 2, the recommended maximum loading rate for an anaerobic lagoon in central South Carolina is 84 g VS/day/m^3 (525 lb VS/day/1000 ft^3). Table 1 shows the loading rate multiplier for swine is 1.0.

$$\text{Design volume} = \frac{285,500}{84\ (1.0)}$$
$$= 3400\ \text{m}^3\ (118,800\ \text{ft}^3)$$

Surface area and depth will be determined by site characteristics. This example uses the maximum recommended loading rate. A lesser loading rate, which would result in a larger lagoon, may be preferable where odor reduction and greater storage capacity is desirable.

4.4.2. Depth — For installations of less than 300 m^3, design depth should be at least 1.5 m (5 ft); greater depths are desirable (see paragraph 4.5.2). Where soil and groundwater conditions permit, depths of 6 m (20 ft) or more may be used. The design surface level should be at least 0.5 m (1.5 ft) above the highest level of the groundwater table.

4.4.3. Length-to-width ratio — Desirable length-to-width ratios vary according to detention time. Lagoons that are roughly round or square are favored for detention periods of 100 days or more or where lagoons are pumped down annually. Length-to-width ratios up to 5 to 1 may be used where detention periods are 30 days or less and where overflow to another lagoon cell or other treatment facility occurs daily.

4.4.4. Detention time — Detention periods of several hundred days are common where influent is limited to animal waste and small amounts of water, where evaporation exceeds precipitation, and/or where recycling of supernatant is practiced. Long detention times can lead to buildup of dissolved salts such as Cl^-, NH_4^+, K^+, and Na^+, which inhibit biological action. When total salt concentration of the supernatant is in the range of 2000 to 3000 mg/ℓ the lagoon should be pumped down and, if necessary, fresh water added (see paragraph 4.8).

Short detention periods may result where hydraulic loading rates are high and where large amounts of wash water enter the lagoon. Minimum detention periods of 15 days are recommended for single cell (single impoundment) lagoons.

4.4.5. Multiple-cell lagoons — Multiple-cell lagoons may be used when required or allowed by local conditions and/or regulation.

When cells operate in series, the initial cell receives the waste; the loading rate should not exceed 125% of the recommended maximum loading rate described in paragraph 4.4.1. Initial cells heavily overloaded in series operation result in severe odor problems.

Where multiple cells are operated in parallel, their volume should be determined according to paragraph 4.4.1.

4.5. Earth embankment

4.5.1. Top width — Minimum top width should be 2.5 m (8 ft).

4.5.2. Side slopes — Side slope on the dry wide of the embankment should not be steeper than 2:1. Wet-side slope should not be steeper than 2:1 above the design surface level. However, slopes to be moved should not be steeper than 3:1. Herbicides should be used on slopes steeper than 3:1. Below the design level, side slope should be as steep as possible without sacrificing soil stability. Deep lagoons and steep wet slopes reduce convection heat loss, enhance internal mixing, reduce odor emission, promote anaerobic conditions, minimize weed growth problems, reduce excavation cost, and reduce mosquito production.

4.5.3. Settlement — Embankment elevation should be increased at least 5% during construction to allow for settling.

4.5.4. Freeboard — The top of the settled embankment should be at least 0.5 m (2 ft) above the maximum design surface level. Where overflow is permitted, a spillway with a minimum capacity of 1.5 times the peak daily inflow rate may be installed, 0.3 m (1 ft) above the design level, to protect the embankment in the case of catastrophic or chronic rainfall events or failure of the overflow control system.

4.6. Inlet and outlet

4.6.1. Inlet — Inlet devices should discharge the waste into the lagoon at a point beyond the cut slope of the embankment. Pipes or open channels may be used. The diameter or least dimension of the inlet should be at least 150 mm (6 in.) or at least 200 m (8 in.) for dairy cattle waste. Inlets may be located above or below the lagoon surface. Submerged inlets may plug unless used at least twice daily. Trickle flow should be eliminated during freezing weather. Access to the inlet device should be provided for cleaning.

4.6.2. Outlet — An outflow device with a minimum capacity of 1.5 times the peak daily inflow rate may be installed at the lagoon surface level *only if* the overflow is to be contained in another lagoon cell or other treatment facility. Outlet devices should be installed in a manner that allows effluent to be taken at a level 150 to 450 mm (6 to 18 in.) below the surface.

4.7. Effluent or overflow

4.7.1. Discharge — Effluent from an anaerobic lagoon may not be discharged to streams, lakes, or other waterways or be allowed to run off the owner's property. Lagoon overflow into a subsequent storage or treatment cell is permissible.

4.7.2. Handling — Lagoon supernatant and sludge are typically spread on land. Sprinkler or surface irrigation, tank spreaders, or other approved distribution procedures are often used. Supernatant should be applied to the land without producing surface runoff, in such amounts that undesirable levels of nutrients or toxic materials do not accumulate in the soil, plants, soil water, groundwater, or runoff, and with procedures that minimize odors. Continuous flow to a surface irrigation system should be avoided; biweekly or weekly dosing is preferable. Removal frequency may be dictated by availability of labor or a suitable application site, cropping schedules, rainfall, or soil conditions.

4.8. Water supply — Adequate water should be available to establish and maintain the minimum operating level of the lagoon at 60% of the design depth. Surface water may be diverted to the lagoon during the start-up period, but must be diverted away after the design depth has been established. The water supply may also be used to dilute lagoon contents and limit salt buildup.

4.9. Safety — A stock-tight fence should enclose the lagoon and warning signs posted to prevent children and others from using the lagoon for other purposes. Fences should be constructed to protect the embankment and to permit access for lagoon maintenance. See ASAE Engineering Practice EP250, Specifications for Farm Fence Construction.

4.10. Vegetation — Trees and shrubs should be cleared for a distance of at least 9 m (30 ft) surrounding the lagoon. This area and dry embankments should be maintained in grass or other low-growing vegetation for erosion prevention and appearance.

5. OPERATION AND MAINTENANCE

5.1. Startup — The water level should be at least 60% of the design depth before waste is added. Such water may come from nearby lakes, streams, ponds, or wells or directed surface runoff or roof drainage. Where possible, waste should be added slowly at first and increased over a period of 2 to 4 months to a design loading rate. Start-up in warm weather is more desirable than in cold weather, particularly in northern states.

5.2. Depth — A water depth of at least 60% of design depth should be maintained to minimize odor production. Daily loading is more satisfactory than less frequent loading. Manure solids should be covered by liquid at all times except as noted below.

5.3. Crust — Floating solids, especially where forage is fed, commonly cause crusting. The crust reduces odor emission and helps maintain anaerobic conditions and constant temperature.

5.4. Sludge removal — Removal of lagoon sludge after many years of operation should be anticipated. Procedures and equipment for management of the sludge are beyond the scope of this document.

5.5. Inspection — Lagoons should be inspected periodically, particularly with regard to operating level and control of overflow. Maintenance is necessary to control rodent damage and growth of vegetation on embankments and surrounding areas.

CITED STANDARDS

1. ASAE EP250, Specifications for Farm Fence Construction.
2. ASAE EP379, Control of Manure Odors.
3. ASAE D384, Manure Production and Characteristics.

SOLIDS SEPARATION UNITS

L. M. Safley

INTRODUCTION

Solids separation devices are components of liquid animal manure handling systems that remove solids for the purposes of refeeding, reuse as bedding, or to reduce the biological load on a treatment facility such as an anaerobic lagoon. Many devices have been suggested for use in separating solids from animal manure slurries, including settling tanks,[1-6] settling channels,[1] vacuum filters,[1,2,8] centrifuges,[1,3,9] troughs using gravity flow of liquids,[7,17] expression processes,[10] stationary screens,[5,11-13] vibrating screens,[2,5,12,14] rotating flighted cylinders,[15] rotating conical screens,[16] and liquid cyclone separators.[12] Of these, stationary inclined screens, vibrating screens, and gravity settling devices are most frequently employed.

TYPES OF DEVICES

Stationary Inclined Screen

A stationary inclined screen separates solids from liquid manure by passing the manure over a sloping screen where the liquids move through the screen and the strained solids are collected at the base. Screens can be single sloped or multisloped and consist of transversely oriented stainless steel bars typically spaced with 0.03- to 0.15-cm openings between bars. Table 1 provides data on the solids removal efficiencies of stationary inclined screens.

Vibrating Screen

A vibrating screen separation device consists of two round trays, one located on top of the other. The top tray has a screen as its base and the lower tray has a solid bottom. As the trays are simultaneously vibrated, the liquid fraction of the slurry passes through the top tray to the lower tray and is discarded, while the solids are retained on the upper tray and subsequently discharged. Table 2 provides data concerning solids removal efficiencies using the vibrating screen.

Gravity Settling Tanks

The manure slurry flows into a sedimentation tank or channel where the heavier solids settle to the bottom and the resulting liquid fraction overflows into storage or treatment facilities. Moore et al.[6] have determined that the settling efficiency is dependent on the retention time in the settling device and the solids concentraiton of the influent slurry. Table 3 presents data concerning settling characteristics of manure from different animal species. Figure 1 indicates the effect of solids concentration on the settling characteristics of swine slurry.

Also important in settling tank design is the rate of overflow and its effect on the removal efficiency. Figure 2 presents this relationship for swine slurry.

Table 4 provides data concerning overflow rates and unit area requirements for settling tanks to achieve an 8% total solids slurry concentration.

Table 1
REMOVAL EFFICIENCY DATA FOR STATIONARY INCLINED SCREENS

Type of screen	Screen opening size (cm)	Slurry passing through screen	Flow rate (ℓ/min/m² of screen)	Solids content of solids removed (%)	Quantity of parameter[a] retained by screen (% of inflow)					Ref.
					TS	TVS	COD	BOD		
Triple sloped	0.15	Dairy slurry[b] (20:1 — water to wet manure by weight)	—	9.4	34	37	—	—		13
Triple sloped	0.15	Dairy slurry[b] (10:1 — water to wet manure by weight)	—	8.4	50	53	—	—		13
Triple sloped	0.15	Dairy slurry[b] (5:1 — water to wet manure by weight)	—	8.7	49	61	—	—		13
Triple sloped	0.08	Dairy slurry[b] (20:1 — water to wet manure by weight)	—	9.5	49	51	—	—		13
Triple sloped	0.08	Dairy slurry[b] (10:1 — water to wet manure by weight)	—	8.0	57	59	—	—		13
Triple sloped	0.08	Dairy slurry[b] (5:1 — water to wet manure by weight)	—	8.5	78	79	—	—		13
Triple sloped	0.03	Dairy slurry[b] (20:1 — water to wet manure by weight)	—	8.1	52	56	—	—		13

Triple sloped	0.03	Dairy slurry[b] (10:1 — water to wet manure by weight)	—	7.5	63	61	—	—	13
Triple sloped	0.03	Dairy slurry[c] (5:1 — water to wet manure by weight)	—	7.7	63	66	—	—	13
Single sloped	0.05	Dairy slurry[b] (7:1 — water to wet manure by weight)	—	6.0	68	67	—	—	13
Single sloped	0.05	Dairy slurry[b] (25:1 — water to wet manure by weight)	—	7.9	63	68	—	—	13
Triple sloped	0.10	Slurry flushed[c] from swine finishing house	352	9.1	35	22	69	62	5
Triple sloped	0.10	Slurry flushed[c] from swine finishing house	526	7.2	26	30	64	38	5
Triple sloped	0.10	Slurry flushed[c] from swine finishing house	675	7.6	28	34	71	52	5
Triple sloped	0.10	Slurry flushed[c] from swine finishing house	897	6.9	11	14	52	14	5
Triple sloped	0.15	Slurry flushed[c] from swine finishing house	352	10.9	3	18	11	14	5
Triple sloped	0.15	Slurry flushed[c] from swine finishing house	526	6.4	4	1	12	4	5
Triple sloped	0.15	Slurry flushed[c] from swine finishing house	675	6.0	10	5	24	34	5

Table 1 (continued)

REMOVAL EFFICIENCY DATA FOR STATIONARY INCLINED SCREENS

Type of screen	Screen opening size (cm)	Slurry passing through screen	Flow rate (ℓ/min/m² of screen)	Solids content of solids removed (%)	Quantity of parameter[a] retained by screen (% of inflow)				Ref.
					TS	TVS	COD	BOD	
Triple sloped	0.15	Slurry flushed[c] from swine finishing house	897	6.0	4	6	24	6	5
Single slope	0.05	Dairy slurry[b] (5.5:1 — water to wet manure by weight)	—	9.5	55—74	57—75	42—63	41—68	11

[a] TS = total solids; TVS = total volatile solids; COD = chemical oxygen demand; BOD = biological oxygen demand.

[b] Low — bedding manure from a stanchion barn mixed with water.

[c] Slurry ranged from 0.2 to 0.7% total solids.

Table 2
REMOVAL EFFICIENCY DATA FOR VIBRATING SCREENS[a]

Screen opening size (cm)	Source of manure[a]	Flow rate (ℓ/min/m² of screen)	Solids content of solids removed (%)	Quantity of parameter[b] retained by screen (% of inflow)			
				TS	TVS	COD	BOD
0.012		248	5.7	2.5	3.3	—	—
0.012		411	8.5	13.8	18.5	—	—
0.012		670	4.8	18.7	42.8	—	—
0.017		248	3.9	5.8	8.4	3.3	—
0.017		411	10.9	14.0	17.0	15.3	—
0.017		670	1.9	1.8	2.4	12.5	—
0.021		248	8.7	14.3	12.4	16.3	4.5
0.021		411	10.8	9.8	12.9	8.9	2.4
0.021		670	4.8	7.0	9.0	10.7	3.9
0.039		248	12.2	12.6	4.3	10.0	—
0.039		411	16.4	22.2	28.1	16.1	—
0.039		670	4.9	12.3	17.2	12.2	—

[a] Manure was a slurry flushed from a swine finishing house. Total solids ranged from 0.2 to 0.7%.
[b] TS = total solids; TVS = total volatile solids; COD = chemical oxygen demand; BOD = biological oxygen demand.

From Shutt, J. W., White, R. K., Taiganides, E. P., and Mote, C. R., *Proc. Int. Symp. Livestock Wastes,* 1975, 463. With permission.

Table 3
TOTAL SOLIDS AND COD REMOVAL (% OF INITIAL) FROM MANURE SLURRIES USING GRAVITY SETTLING TANKS

Animal	Solids concentration (%)	Settling time (min)							
		1		10		100		1000	
		TS[a]	COD[a]	TS	COD	TS	COD	TS	COD
	1.0	40	23	61	45	67	54	71	57
Beef	0.1	46	41	59	57	63	63	64	67
	0.01	27	41	51	51	59	57	66	63
	1.0	50	46	64	64	68	71	70	73
Dairy	0.1	50	35	61	45	65	50	67	54
	0.01	44	57	51	62	57	67	62	71
	1.0	65	58	76	67	79	71	81	72
Horse	0.1	57	42	66	59	72	68	77	72
	0.01	64	51	70	62	74	66	77	70
	1.0	53	50	71	66	75	68	76	69
Poultry	0.1	55	42	69	58	73	66	75	70
	0.01	43	29	64	43	74	50	75	55
	1.0	52	26	62	33	66	40	70	46
Swine	0.1	44	20	55	28	62	31	68	35
	0.01	36	46	45	56	52	63	59	70

Note: Samples were collected at the top of cylinder at time intervals of 1, 10, 100, and 1000 min and compared to initial material.

[a] TS = total solids; COD = chemical oxygen demand.

From Moore, J. A., Hegg, R. O., Scholz, D. C., and Strauman, E., Settling Solids in Animal Waste Slurries, ASAE Paper 73-438, American Society of Agricultural Engineers, St. Joseph, Mich., 1973. With permission.

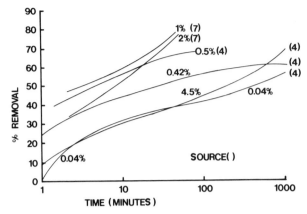

FIGURE 1.   Comparison of the effect of solids concentration on the settling characteristics of swine manure slurry.

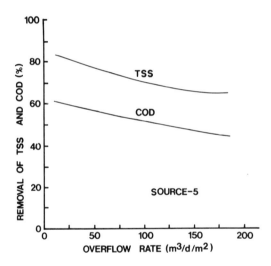

FIGURE 2.   Total suspended solids (TSS) and chemical oxygen demand (COD) removal efficiencies at various overflow rates (cubic meters per day per square meter) for a slurry with initial TSS concentration of 5000 mg/ℓ.

Table 4
**UNIT AREA AND OVERFLOW RATE
REQUIREMENTS IN A SETTLING CHAMBER TO
CONCENTRATE SOLIDS TO 8% FOR SLURRIES
WITH VARIOUS INITIAL SOLIDS
CONCENTRATIONS**

Initial total solids concentration		Unit area required (m^2/ton/day)	Overflow rate (m^3/day/m^2)
%	mg/ℓ		
1	10,000	1.1	89
2	20,000	3.0	17
4	40,000	6.6	4

From Shutt, J. W., White, R. K., Taiganides, E. P., and Mote, C. R., *Proc. Int. Symp. Livestock Wastes,* 1975, 463. With permission.

REFERENCES

1. **Pratt, G. L.,** System Components to Separate Solids and Liquids, Animal Waste Management Conference, Ames, Iowa, October 13 to 15, 1971.
2. **Glerum, J. C., Klomp, G., and Poelma, H. R.,** The separation of solid and liquid parts of pig slurry, *Proc. Int. Symp. Livestock Wastes,* 1971, 345.
3. **Holmes, L. W. J., Day. D. L., and Pfefer, J. T.,** Concentration of proteinaceous solids from oxidation ditch mixed-liquor, *Proc. Int. Symp. Livestock Wastes,* 1971, 351.
4. **Fischer, J. R., Sievers, D. M., and Fulhage, C. D.,** Settling characteristics of swine manure as related to digestor loading, *Proc. Int. Symp. Livestock Wastes,* 1975, 456.
5. **Shutt, J. W., White, R. K., Taiganides, E. P., and Mote, C. R.,** Evaluation of solids separation devices, *Proc. Int. Symp. Livestock Wastes,* 1975, 463.
6. **Moore, J. A., Hegg, R. O., Scholz, D. C., and Strauman, E.,** Settling Solids in Animal Waste Slurries, ASAE Paper 73-438, American Society of Agricultural Engineers, St. Joseph, Mich., 1973.
7. **Jett, S. C., Ross, I. J., Hamilton, H. E., and Hays, V. W.,** Size distribution and nutritional value of swine manure separates, *Trans. ASAE,* 17(5), 965, 1974.
8. **Backer, L. F., Witz, R. L., Pratt, G. L., and Buchanan, M. L.,** Vacuum Filtration of Cattle Manure, ASAE Paper 73-4531, American Society of Agricultural Engineers, St. Joseph, Mich., 1973.
9. **Ross, I. J., Begin, J. J., and Midden, T. M.,** Dewatering poultry manure by centrifugation, *Proc. Int. Symp. Livestock Wastes,* 1971, 348.
10. **Steffe, J. F. and Gerrish, J. B.,** Dewatering a Swine Manure Slurry by Expression, ASAE Paper 77-4577, American Society of Agricultural Engineers, St. Joseph, Mich., 1977.
11. **Graves, R. E., Clayton, J. T., and Light, R. G.,** Renovation and reuse of water for dilution and hydraulic transport of dairy cattle manure, *Proc. Int. Symp. Livestock Wastes,* 1971, 341.
12. **Bartlett, H. D., Bos, R. E., and Wanz, E. C.,** Dewatering bovine manure, *Trans. ASAE,* 17(5), 968, 1974.
13. **Graves, R. E. and Clayton, J. T.,** Stationary Sloping Screen to Separate Solids from Dairy Cattle Manure Slurries, ASAE Paper 72-951, American Society of Agricultural Engineers, St. Joseph, Mich.,
14. **Fairbank, W. C. and Bramhall, E. L.,** Dairy Manure Liquid-Solids Separation, Univ. Calif. Agric. Ext. Ser. Pub. AXT-271, University of California, 1968.
15. **Miner, J. R. and Verley, W. E.,** Application of the rotating flighted cylinder to livestock waste management, *Proc. Int. Symp. Livestock Wastes,* 1975, 452.
16. **Shirley, R. and Butchbaker, A.,** A rotating conical screen separator for liquid-solid separation of beef waste, *Proc. Int. Symp. Livestock Wastes,* 1975, 459.
17. **Pratt, G. L., Buchanan, M. L., and Witz, R. L.,** Cable Driven Scrapers for Manure Collection and Liquid-Solid Separation, ASAE Paper 74-4003, American Society of Agricultural Engineers, St. Joseph, Mich., 1974.

PRINCIPLES AND DESIGN OF INFILTRATION SYSTEMS

D. H. Vanderholm

INTRODUCTION

Complete infiltration into agricultural soils is a satisfactory treatment method for many types of livestock wastes. Systems of this type are commonly referred to as infiltration systems, but this may vary depending upon configuration and locality. The wastes are normally surface applied, as subsurface infiltration has generally been unsuccessful for livestock wastes. Infiltration systems differ from conventional waste irrigation systems in that the land is devoted specifically to waste treatment. Although a crop may be grown on the treatment area, and this is desirable, the primary objective is to have an effective and economical waste treatment system. The fertilizer nutrient value and the irrigation water benefits are of secondary importance.

Treatment of food processing and municipal waste water by infiltration has risen in popularity over recent years. Some of these are designed for total infiltration while others are more properly termed irrigation disposal. There are numerous reports and planning manuals dealing with these infiltration systems and some are included in the list of references.[1-5]

INFILTRATION FOR TREATMENT OF LIVESTOCK WASTES

Total infiltration has not been widely used for treating livestock wastes, but it does offer an alternative that may be attractive in certain situations, especially those in which land for application is limited. Various configurations have been studied and reported.[6-8] Infiltration is often an important treatment phase in overland flow treatment systems, but the objective with these is to treat and surface discharge the waste. In contrast, properly designed and managed infiltration systems would not be expected to discharge wastes except under extreme circumstances.

Infiltration is normally only satisfactory for relatively dilute liquid wastes with low solids contents. Examples of these would be open feedlot runoff and treatment lagoon effluent. Wastes with high solid content will cause soil clogging and greatly inhibit infiltration and plant growth. Generally speaking, solids content of wastes to be infiltrated should not exceed 0.5%. Settling or other solids removal techniques should be used for open feedlot runoff and other wastes, which may have higher solids contents, before the waste is applied to the infiltration area.

The use of a soil mantle in wastewater renovation is reviewed in detail by McGaughey and Krone.[9] For best pollutant removal, it is desirable to maintain the soil in an aerobic condition as much as possible. This can be accomplished by having adequate drainage and by resting the infiltration area sufficiently between waste applications. Intermittent application also aids infiltration rate recovery, which tends to be reduced during waste application.

Soil filtration is effective in pollutant removal, with observed reductions of 90 to 95% of applied chemical oxygen demand and phosphorous and 50 to 60% of applied nitrogen during percolation through a depth roughly corresponding to the root zone. Nitrification occurs with periodically loaded systems and some of the resulting nitrate is leached through the system and into the groundwater. This can be a concern in heavily loaded systems and monitoring of groundwater nitrate levels may be advisable so that changes in management can be instituted if necessary to prevent excessive groundwater nitrate levels from occurring.

Denitrification in wet surface layers of the soil can result in significant nitrogen losses to the atmosphere that can be beneficial in reducing the hazard to groundwater.

DESIGN

Infiltration systems are normally designed on a hydraulic loading basis. Independent design variables include volume of wastewater and its pattern of occurrence, soil type, topography, and climate. Wastewater storage is usually necessary to allow planned application even with varying wastewater flows. The structures may be very simple such as level infiltration areas surrounded with earth berms to catch and store storm runoff until it infiltrates; or large ponds with capacity for several months waste accumulation. The larger storage would be required in areas where soil infiltration rates are adversely affected by the climate. Ideally, the storage capacity should be sufficient to avoid applying wastewater to the infiltration area during frozen or saturated soil conditions unless discharge from the land is acceptable.

In some locations, increased nitrate levels in groundwater may be of concern, and application rates may be governed by maximum nitrogen loading rather than maximum hydraulic loading. Increased soil salinity and accumulations of phosphorus can also be of concern. With infiltration systems for livestock wastes, applied nutrients will usually be far in excess of crop nutrient needs. Whether or not this is a hazard depends upon the individual situation. A thorough literature review on effects of high application rates was prepared by Powers et al.[10] Local regulations that might limit application rates should also be checked prior to planning an infiltration system.

The basic steps for designing an infiltration system can be summarized as follows:

1. Determine waste volume and pattern of occurrence. For storm runoff from open lots, a design storm size must be selected and a runoff quantity predicted. For systems with a known effluent output such as manure and flushing water, a relatively accurate waste flow can be calculated.

2. Determine infiltration characteristics of the soil. This can be done by determining the soil type (either from soil maps or on-site inspection) and using irrigation guides that give normal infiltration ratings for that soil. There are usually available from Cooperative Extension or Soil Conservation Service offices. Wastewater does not infiltrate as rapidly as clean water in most soils; therefore, it is usually advisable to reduce clean water infiltration rates by 50 to 75% when designing waste systems.[11] For larger systems, actual field infiltration data should be obtained using infiltrometers. Table 1 has been included as a rough guide when specific information is not available.

3. Select the application frequency and quantity of waste to be infiltrated at each application. This design is not applicable for treatment of storm runoff unless additional storage is provided. With no storage, the quantity will be the runoff volume applied when the storm occurs. For continuously discharged effluent, a volume and frequency is selected for the waste to be removed from storage and applied to the infiltration area. Continuous application is not recommended because no soil recovery time is provided.

4. Select a time period to infiltrate the waste quantity per application and calculate the total area required to accomplish this.

5. Design an application system to apply waste at the approximate rate. If the infiltration area is contained within berms and short-term ponding is allowed, a regulated application rate is unnecessary. In other situations where the infiltration area is not contained or where agronomic factors limit ponding, the application rate should not exceed infiltration rate to prevent excessive surface flooding and runoff. Application may be made by sprinkler or surface irrigation equipment, gravity flow, or other suitable methods.

Table 1
SUGGESTED MAXIMUM WASTE
APPLICATION RATES[a]

	mm/hr	
Soil characteristics	Cover	Bare
Clay, poorly drained	7.6	3.8
Silty surface, tight, poorly drained subsoil	10.12	6.4
Silt loam, well drained	15.2	10.2
Sandy loam, well drained	23.0	15.2

[a]   For ground with 5% slope or less. Rates should be reduced for greater slopes.

6. Consider possible environmental hazards associated with the system and take necessary precautions in system design, monitoring, and operating procedures to minimize potential hazards. These would include excessive nutrient loading, odor emissions, spray drift, disease, and toxic compounds accumulated in the crop.
7. Specify cropping and management procedures for the infiltration area. It is desirable to remove as much of the applied nutrients as possible through cropping to minimize leaching and soil accumulation. Hay production adapts well to this situation. Grazing may also be permitted, but grazing or harvesting when the soil is wet should not be allowed as soil compaction, rough surfaces (resulting in surface ponding), and reduced infiltration will result.

The following two examples serve to illustrate the design procedure for infiltration systems.

Example 1

An infiltration system is to be used to treat storm runoff from an open beef feedlot, earth surface, 2 ha in size. The soil in the proposed infiltration area is a silty clay loam with an expected infiltration rate of 5 mm/hr.

Runoff will flow by gravity through a settling channel for solids removal and then to the infiltration area. A flat infiltration area with surrounding earth berms to provide temporary storage for the entire design storm runoff will be used.

The system will be designed for a 10-year, 24-hr storm of 125 mm of which 90% or 112.5 mm is expected to reach the infiltration system as runoff. Total volume of runoff then is 2(112.5) = 225 ha/mm. It is desired to infiltrate the runoff plus direct precipitation on the infiltration area within 24 hr following the runoff event. Select a trial area of 2 ha, equal to the feedlot area. Direct precipitation plus runoff to be applied is 2(125) + 225 = ha/mm. With this area, 475 ÷ 2 = 237.5 mm must be infiltrated in 48 hr (24-hr storm duration and 24-hr infiltration time). At 5 mm/hr, 240 mm can be infiltrated in 48 hr; thus, this area is adequate. An earth berm high enough to store the depth to be infiltrated during the second 24-hr period should be constructed around the infiltration area. In this situation, the depth

Table 2
DESIGN OF INFILTRATION AREA FOR 10-YEAR, 25-HR HYPOTHETICAL STORM

Direct rainfall on infiltration area	125 mm
Added feedlot runoff = (2 × 125) 0.9	225 ha/mm
Total water on infiltration area = 225 + (125 × infiltration area, ha)	475 ha/mm
Infiltration during 24-hr storm (0.5 × 240)2	240 ha/mm
Volume ponded after 24-hr storm	235 ha/mm
Depth ponded after 24 hr storm 235 ÷ 2	117.5 mm
Freeboard depth	150 mm
Impoundment depth required (berm height)	268 mm

needed would be $237.5 - 120 = 117.5$ mm plus 100 to 150 mm freeboard. These design steps are summarized again in Table 2.

Example 2

An infiltration system is to be used for disposal of effluent from an anaerobic lagoon system treating the water from a flushing gutter, swine finishing building. Lagoon effluent is recycled for flush water. Building capacity is 500 animals with an average weight of 60 kg. The lagoon effluent will be applied by an irrigation sprinkler on a sandy loam area with a slope of 2%.

There is not net loss or gain to the lagoon system from direct precipitation, evaporation, or seepage; therefore, the total volume of waste to be applied to the infiltration area is equivalent to the volume of manure, spilled feed, and water spillage from the hog facility. The average volume of waste produced per hog is 3.75 ℓ daily. Increasing this by 20% for feed and water spillage, the total volume of waste to treat annually is $1.2 \times 3.75 \times 500 \times 365 = 821,250$ ℓ.

Due to winter conditions, no waste will be applied for 5 months; therefore, the total waste will be applied at regular intervals over a 7-month period. Assume that the total application to the infiltration area should not exceed 100 mm/year so that nutrient applications are not excessive. In the case of a strong waste such as swine lagoon effluent, nutrient applications may be the limiting factor, rather than hydraulic loading. Where this is not a concern or if a very dilute waste is being applied, total applications may be considerably higher without damaging crop growth and infiltration rates.

Using a 100-mm maximum, one million liters can be applied to each hectare, so that total area required is $821,750 \div 1,000,000 = 0.82$ ha.

The waste can be applied in four separate applications of 25 mm at approximately 2-month intervals. This can easily be accomplished with small, hand-moved portable irrigation equipment.

These examples illustrate the general procedure for planning an infiltration system. Due to extreme variability in waste characteristics, local conditions, regulations, and other factors, general loading rates and designs cannot be recommended. The designer of an infiltration system must determine this information for the specific situation and plan accordingly. It is essential that adequate preliminary investigation be carried out, in order to plan a system within the restraints of normal farm operation that is manageable and environmentally acceptable. Infiltration can provide an economical, low-maintenance treatment system, but only if planned, installed, and managed properly.

REFERENCES

1. **Thomas, R. E.**, Land treatment of wastewater — an overview of methods, *J. Water Pollut. Control Fed.*, 45, 1476, 1973.
2. National Canners Association, Liquid Wastes from Cannery and Freezing Fruits and Vegetables, Final Rep. Proj. 12060 EDK, U.S. Environmental Protection Agency, Washington D.C., 1971.
3. **Parizek, R. R., Kardos, L. T., Sopper, W. E., Myers, E. A., Davis, D. E., Farell, M. A., and Nesbitt, V. B.**, Wastewater Renovation and Conservation, Study, 32, Pennsylvania State University, University Park, 1967.
4. The Irrigation Association, *Wastewater Resource Manual*, Sprinkler Irrigation Association, Silver Spring, Md., 1975.
5. Cold Regions Research and Engineering Laboratory, Wastewater Management by Disposal on the Land, Rep. AD-752 132, National Technical Information Service, Springfield, Va., 1972.
6. **Koelliker, J. K., Miner, J. R., Beer, C. E., and Hazen, T. B.**, Treatment of livestock effluent by soil filtration, in *Livestock Waste Management and Pollution Abatement*, ASAE Proc. 271, American Society of Agricultural Engineers, St. Joseph, Mich., 1971, 329.
7. **Swanson, N. P., Linderman, C. L., and Mielke, L. N.**, Direct land disposal of feedlot runoff, in *Managing Livestock Wastes*, ASAE Proc. 275, American Society of Agricultural Engineers, St. Joseph, Mich., 1975, 255.
8. **Ritter, W. F. and Eastburn, R. P.**, Treatment of livestock wastes by a barrier landscape water renovation system, in *Managing Livestock Wastes*, ASAE Proc. 275, American Society of Agricultural Engineering, St. Joseph, Mich., 1975, 572.
9. **McGauchey, P. H. and Krone, R. B.**, Soil Mantle as a Wastewater Treatment System, Rep. 67-11, San. Eng. Res. Lab., University of California, Berkeley, 1967.
10. **Powers, W. L., Wallingford, G. W., and Murphy, L. S.**, Research Status on Land Application of Animal Wastes, Report EPA 660/2-75-010, U.S. Environmental Protection Agency, Washington, D.C., 1975.
11. **Vanderholm, D. H. and Beer, C. E.**, Use of the soil to treat anaerobic lagoon effluent: design and operation of a field disposal system, *Trans. ASAE*, 13, 562, 1970.

VEGETATIVE INFILTRATION AREAS

John C. Nye

INTRODUCTION

One technique to control feedlot runoff is vegetative infiltration areas. Runoff that has passed through a settling basin is allowed to flow through this vegetative infiltration area where it is filtered, degraded, or infiltrated into the soil and vegetation and prevented from polluting any water course. Numerous techniques to evaluate these runoff control systems have been used. The most important criteria appear to be (1) infiltration rate, (2) travel time, and (3) nutrient loading.

INFILTRATION RATE TECHNIQUE

Initial procedures for design of vegetative infiltration areas considered infiltration rate as the primary criteria. Research by Vanderholm et al.[12] confirms the importance of this parameter. The general procedure is to determine the total volume of runoff and rainwater entering the infiltration area. The infiltration area is then determined as follows.

$$V_I = I \times A_i + Q \times A_f$$

Simplify by assuming all rainfall on feedlot results in runoff during design storm.

$$V_I = I \times (A_I + A_i)$$

also

$$f = (V_I)/(t \times A_I)$$

or

$$A_I = (A_f * I)/(f - I)$$

where V_I = volume of runoff that must be infiltrated, I = rainfall, A_i = area of infiltration area, Q = volume of runoff from feedlot, A_f = area of feedlot, f = infiltration rate of soil, and t = duration of storm.

The infiltration rates for soils are dependent on the moisture content in the soil, but average values have been used to design infiltration areas. These values are presented in Table 1.

A critical parameter in sizing the vegetative infiltration area is the selection of the design storm duration. Vanderholm et al.[12] suggested that the 1-year, 2-hr storm would be the design storm of Illinois conditions. This was based on the research conducted at four sites which indicated that 2-hr contact time in the vegetative infiltration areas would be adequate to remove more than 95% of the pollutants from the feedlot runoff.

There are numerous uncertainties in using infiltration rates to design a vegetative infiltration area. The infiltration rate depends on the moisture content and temperature of the soil. An exact performance cannot be predicted, but experience shows that systems can reduce the pollutants in feedlot runoff and the quantity of water leaving the vegetative infiltration area is substantially reduced to provide better control of the runoff discharge as a second parameter; travel time should be determined.

Table 1 INFILTRATION RATE OF VARIOUS SOILS		
	Infiltration rate	
Soil	**mm/hr**	**in/hr**
Silty clay loam	30.8	1.2
Silt loam	38.0	1.5
Sandy loam	43.0	1.7

Table 2
MINIMUM CONTACT TIME FOR VEGETATIVE INFILTRATION AREAS WHERE CHANNELIZED FLOW IS USED

Lot size		**Travel time (hr)**
m²	**ft²**	
929	10,000	2
1394	15,000	3
1858	20,000	4
2323	25,000	5

TRAVEL TIME

The time required to filter out, absorb, or infiltrate pollutants has been estimated by Vanderholm et al.[12] They divided vegetative infiltration areas into two categories: overland flow and channelized flow. Two-hour contact time is required in the overland flow systems. For channelized flow, the contact time is a function of feedlot size as shown in Table 2.

The contact time can be determined using Manning's equation with a roughness coefficient of 0.3 according to Vanderholm et al.[12] Depth of flow is assumed to be 1.3 cm. According to Kao and Barfield,[4] Manning's equation should not be applied to shallow flow in dense vegetation. They found that under nonsubmerged flow conditions, the major force controlling flow was the drag force from the vegetative blade. For shallow flow, the velocity will be less than that predicted by Manning's equation.

NUTRIENT LOAD

A third criteria of importance is the nutrient load on the vegetative infiltration area. Most of the pollutants in the feedlot runoff that have passed through a settling basin will be in the form of dissolved nutrients. These nutrients must be assimilated into the environment. Bingham et al.[1] found that 90 to 100% removal of Kjeldahl nitrogen, phosphorous, nitrate, TOC, chemical oxygen demand, and chlorides was possible using a grass buffer strip around an area that received poultry manure. Overcash et al.[8] have developed a mathematical model to predict the reduction in pollutants concentration as a function of the infiltration rate to rainfall rate ratio and the buffer area to waste area length ratio. They concluded that the vegetative infiltration area width should be equal to the waste area length for clay loam soils if a >90% reduction in concentration was required.

In other work involving the application of municipal sewage effluent to grass filtration areas, Butler et al.[2] reported that a contact time of 6 hr is required to reduce the nitrate concentration to an acceptable level.

Nye and Jones[7] reported that the nutrient load to the vegetative infiltration area should be no more than 50% greater than the crop nutrient uptake. At that nutrient loading, the vegetative infiltration area is approximately the same size as the feedlot.

CONCLUSION

Control of feedlot runoff can be accomplished by numerous techniques. Livestock producers with small feedlots, less than $1/2$ acre, can consider settling basins and vegetative infiltration areas. Larger quantities of runoff usually require the construction of a holding pond to contain the runoff until it can be applied to crop land with irrigation equipment.

The size of the components is dependent on the size of the feedlot, anticipated rainfall and runoff, dewatering system, and management plan. Once these parameters are known, the design of the facilities can be integrated into the overall livestock production complex.

This description of vegetative infiltration areas can be used to design facilities to handle feedlot runoff that has passed through a primary treatment, such as a settling basin, to remove suspended solids.

REFERENCES

1. **Bingham, S. C. Overcash, M. R., and Westerman, P. W.,** Effectiveness of Grass Buffer Zones in Eliminating Pollutants in Runoff from Waste Application Sites, ASAE Paper No. 74-2571, American Society of Agricultural Engineers, St. Joseph, Mich., 1978.
2. **Butler, R. M., Myers, E. A., Walter, E. A., and Husted, J. V.,** Nutrient Reduction in Wastewater by Grass Filtration, ASAE Paper No. 74-4024, American Society of Agricultural Engineers, St. Joseph, Mich., 1974.
3. **Gilbertston, C. B., Clark, N., Nye, J., and Swanson, N.,** Runoff Control for Livestock Feedlots — State of the Art, ASAE Paper No. 79-4071, American Society of Agricultural Engineers, St. Joseph, Mich., 1979.
4. **Kao, D. T. Y and Barfield, B. J.,** Prediction of flow hydraulics for vegetated channels, *Trans. ASAE,* 21, 489, 1979.
5. **Norman, D. A., Edwards, W. M., and Owens, L. B.,** Design Criteria for Grass Filter Areas, ASAE Paper No. 78-2573, American Society of Agricultural Engineers, St. Joseph, Mich., 1978.
6. **Nye, J. C., Jones, D. D., and Sutton, A. L.,** Runoff Control Systems for Open Livestock Feedlots, ID 114, Cooperative Extension Service, Purdue University, West Lafayette, Ind., 1976.
7. **Nye, J. C. and Jones, D. D.,** Experiences with runoff control systems in Indiana. *Trans. ASAE,* 23, In press.
8. **Overcash, M. R., Humenik, F. J., Westerman, P. W., Covil, D. M., and Gillian, J. W.,** Overland Flow Pretreatment of Poultry Manure, ASAE Paper No. 76-4517, American Society of Agricultural Engineers, St. Joseph, Mich., 1978.
9. **Sutton, A. L., Jones, D. D., and Brumm, M. C.,** A Low Cost Settling Basin and Infiltration Channel for Controlling Runoff from an Open Swine Feedlot, ASAE Paper No. 76-4516, American Society of Agricultural Engineers, St. Joseph, Mich., 1976.
10. **Swanson, N. P., Linderman, C. L., and Mielke, L. N.,** Direct land disposal of feedlot runoff, in *Managing Livestock Wastes,* American Society of Agricultural Engineers, St. Joseph, Mich., 1975, 255.
11. **Vanderholm, D. H. and Dickey, E. C.,** Design of Vegetative Filters for Feedlot Runoff Treatment in Humid Areas, ASAE Paper No. 78-2570, American Society of Agricultural Engineers, St. Joseph, Mich., 1978.
12. **Vanderholm, D. H., Dickey, E. C., Jackobs, J. A., and Elmore, R. W., Spahr, S. L.,** Livestock Feedlot Runoff Control by Vegetative Filters. EPA-600/2-79-743, National Technical Information Service, Springfield, Va., 1979.
13. **Vanderholm, D. H. and Nye, J. C.,** Systems for Runoff Control, PIH-21, Cooperative Extension Service, Purdue University, West Lafayette, Ind., 1978.

EQUIPMENT FOR SOLID MANURE HANDLING AND LAND APPLICATION

John M. Sweeten

OCCURRENCE OF SOLID MANURE

Manure having a moisture content of 80% or less can be handled as a solid.[1] Solid manure handling is practiced at the vast majority of cattle feedlots in the Great Plains, midwest, and western sections of the U.S. Many commercial dairies in the southwestern U.S. also utilize open lot feeding of cattle and, consequently, solid manure handling methods. Essentially all sheep feedlots — open lot or confinement building — utilize solid manure handling equipment. Many dairies in the northern and eastern U.S. utilize total confinement barns and/or open lots that require solid manure handling. A high percentage of U.S. poultry production — broilers and caged layers — occurs in buildings designed for solid manure removal and disposal. A relatively small percentage of U.S. hog production occurs on solid concrete floors that require handling of manure as a solid.

CHARACTERISTICS AND AMOUNTS

Generally, manure collected from open lots has a moisture content of 20 to 50%. Manure scraped from floors of confinement buildings contains 50 to 80% moisture, with the exception of broiler litter (manure and bedding material), which usually contains less than 35% moisture. Amounts to be removed vary within animal species depending upon ration, animal spacing, feedlot surfacing material, cleaning frequency and procedures, use of bedding, time in confinement, and other factors.[2]

To illustrate, beef cattle on high concentrate rations excrete about 24 kg of wet manure (85% moisture, wet basis) per day per 385-kg steer. This amounts to 3.6 kg per head per day of manure solids, of which 20% is ash. Additional ash is collected in the form of soil in unsurfaced feedlots. Offsetting this are processes of evaporation and biological decomposition so that an average of 4.9 kg of collectable manure per head per day (at 40% moisture) is harvested from the feedlot surface (Table 1).[1,3]

Climatic conditions and management practices affect manure quality with regard to value as fertilizer, fuel, or feedstuff. Fiber and ash contents increase as decomposition progresses. Nitrogen losses of 50 to 60% can occur on the feedlot surface. Hence, the value of manure is improved by frequent collection.

SOLID MANURE MANAGEMENT STEPS

Management of solid manure involves three functions: collection, loading, and spreading. An intermediate step such as composting and/or stockpiling is sometimes employed. Machines used to collect solid manure include tractor-mounted scraper blade, wheel loader (with or without prior chisel plowing or rotary tilling), track loader, elevating scraper, and/or road grader. Loading is performed by wheel or track loaders or tractor-mounted bucket loaders. Spreader trucks and tractor-towed manure spreaders are used for hauling and distribution.

Manure removal frequencies are dictated by climatic conditions, animal comfort, access to land, labor scheduling, storage capacity, and water and air pollution potentials. In most open feedlots, solid manure is collected about two or three times annually. Annual or semiannual collection is practiced in most poultry houses. By contrast, daily or twice-weekly manure collection occurs in most dairy barns.

Table 1
MANURE REMOVED FROM SOUTH-CENTRAL KANSAS
UNSURFACED CATTLE FEEDLOT PENS

Cleaning date		Feeding period (days)	Animal spacing (m²)	Dry solids removed, daily basis (kg/animal)
Aug.	1969	290[a]	25	4.7
Jan.	1969	287[a]	23	6.0
Nov.	1969	153	20	3.6
Jan.	1970	155	25	6.2
Feb.	1970	141	23	6.5
Feb.	1970	163	24	5.4
Mar.	1970	129	26	4.4
June	1970	138	26	4.9
July	1970	149	25	1.7
July	1970	139	29	2.4
Sep.	1970	146	21	1.2
Nov.	1970	146	23	4.0
Dec.	1970	144	23	6.0
Dec.	1970	147	26	6.8
May	1971	159	26	7.9
May	1971	160	24	7.8

[a] Represents two feeding periods without intermediate cleaning.

From Miner, J. R. and Smith, R. J., Eds., Livestock Waste Management with Pollution Control, North Central Regional Res. Publ. 222, Midwest Plan Service Handbook, MWPS-19, Midwest Plan Service, Ames, Iowa, 1975, 23. With permission.

Factors that influence the efficiency of manure handling include type and size of equipment, operator skill and technique, building layout or pen size, terrain, presence of obstacles, manure moisture content, and operator fatigue.

Butchbaker et al.[4] found that investment and operating costs of solid manure handling systems varied with feedlot size, manure hauling distance, and equipment usage. These costs average 12% of the total operating expense of a 20,000-head feedlot.[4,5]

SOLID MANURE COLLECTION EQUIPMENT

The most commonly used type of solid manure collection unit for concrete lots and alleys is the rear scraper blade, designed to mount on 22- to 120-kW tractors. Tractor-mounted bucket loaders (680- to 1700-kg lift capacity) are widely used for collecting manure from both concrete and dirt surfaced feedlots. At large commercial feedlots, the wheel loader (75 to 130 kW) is the standard piece of machinery used. Often the wheel loader is preceded by chisel plowing or rotary tilling. Small elevating scrapers (100 to 250 kW) are sometimes used for feedlot manure colleciton, assisted by a wheel loader. Machine usage (hr/year) for manure collection at different size feedlots is estimated in Table 2.

Basic systems for feedlot manure collection are wheel loader, wheel loader plus plowing or rotary tilling, elevating scraper, and road grader plus elevating scraper. Manure collection rates for the first three of these systems are compared in Table 3.[2]

A tightly compacted manure surface can be broken up using a tractor-mounted rotary tiller or chisel plow at the rate of 31 ± 3 m²/min.[2] Manure collection rates with a wheel loader are closely related to the amount of manure collected per cycle, or per bucketful. A regression equation ($R^2 = 0.94$) to express this relationship is[2]

Table 2
ESTIMATED ANNUAL MACHINE USAGE (HR/YEAR) FOR MANURE COLLECTION FROM OPEN FEEDLOTS

Cattle feedlot capacity[a] (head)	Annual collectable manure quantity[b] (t/year)	Manure collection rate[c] (t/hr)					
		50	100	150	200	250	300
100	150	3	1.5	1	0.7	—	—
1,000	1,500	30	15	10	7.5	6	5
5,000	7,500	150	75	50	38	30	25
10,000	15,000	300	150	100	75	60	50
50,000	75,000	1,500	750	500	375	300	250
100,000	150,000	—	1,500	1,000	750	600	500

[a] Applies to beef feedlots or open lot dairy corrals; for sheep feedlots, assume 1 head cattle = 6 sheep.
[b] Assumes 5 kg/day collectable manure production (wet basis) and animal occupancy at 80% of feedlot capacity.
[c] Manure collection rate at actual field efficiency, which averages 75 to 80% for wheel loaders and 90% for elevating scrapers.

From Sweeten, J. M., Great Plains Beef Cattle Handbook L-1769, Tex. Agric. Ext. Service, Texas A & M University, College Station, 1979. With permission.

Table 3
COMPARISON OF CERTAIN FEEDLOT MANURE COLLECTION SYSTEMS ACCORDING TO PRODUCTIVITY AND ENERGY REQUIREMENTS

Primary collection machine	Feedlot surface preparation	Manure collection rate at 100% operating efficiency (t/hr)	Energy requirements at 100% operating efficiency (kW-hr/t)
Elevating scraper	None	114	0.88
Wheel loader	Rotary tilled	106	0.99
Wheel loader	None	107	1.28
Wheel loader	Chisel-plowed three times	160	0.96

From Sweeten, J. M. and Reddell, D. L., *Trans. ASAE,* 22, 138, 1979. With permission.

$$CR = 120 \, MC - 25.5 \tag{1}$$

where CR = collection rate (t/hr), and MC = manure collected per cycle, i.e., per bucketful (t).

Conditions and techniques that enable the operator to obtain a "full bucket" each collection cycle will result in high productivity. Chisel plowing the surface before collection reduces particle size, loosens the manure surface, and reduces mechanical energy needed to excavate and otherwise tightly compacted manure pack with the loader. Thus, plowing or rotary tilling the manure pack increases productivity and may also tend to improve manure spreading uniformity and, hence, farmer/customer satisfaction.[2]

Operating efficiencies (ratio of productive time to total elapsed time) determined from a feedlot time-motion study[2,3] ranged from 79% for wheel loaders to 90% for an elevating scraper (Table 4).

Table 4
OBSERVED OPERATING
EFFICIENCY (%) FOR SOLID
MANURE COLLECTION
MACHINERY AT SOUTHERN
GREAT PLAINS CATTLE
FEEDLOT[2,3]

	Percent of operating efficiency	
Machinery	Average	Range
Wheel loaders	79	70—87
Elevating scraper	90	78—98
Tractor-drawn scarifier	84	56—93

EQUIPMENT FOR LOADING MANURE TRUCKS AND TOWED SPREADERS

The most commonly used equipment for loading solid manure into trucks or manure spreaders is the bucket loader mounted on a farm tractor. Sizes of tractors (22 to 120 kW) and loader buckets (565- to 1700 kg lift capacity) vary widely, the choice depending upon pen size, alley width, vertical clearance, amount of manure, and cost.

At commercial feedlots and open lot dairies with more than about 2500 head, a wheel loader is used to load manure into trucks. Wheel loader sizes commonly utilized are 75 to 130 kW with bucket "heaped load" capacities of 1.1 to 3.4 m³ (950 to 2850 kg). A combination of conveyor belt and wheel loader has been used in a few feedlots. In some cases, an elevating scraper has been used to haul manure to a stockpile, where later reloading with a wheel loader is necessary.

At commercial feedlots, measured rates of manure truck loading, using 75- to 108-kW wheel loaders, have averaged 186 t/hr, with a range of 145 to 254 t/hr (adjusted to 100% operating efficiency).[2] Actual operating efficiencies often average only 25% because of time spent waiting for spreader trucks to return. Time required to load a mixture of single- and dual-axle spreader trucks (9.3 per load) averages 3.0 ± 0.7 min. The energy requirement for loading manure range from 0.40 to 0.67 kW-hr/t.[3]

Estimated hours of machine usage per year for manure loading at different size livestock operations are estimated in Table 5.

SOLID MANURE SPREADING EQUIPMENT

According to Miner and Smith,[1] most spreaders used for transporting solid manure are box type with rear discharge. However, a few manufacturers supply side-discharge open-tank spreaders. Besides serving as the transport vehicle for the manure, the spreader should provide for a uniform distribution of the manure on the cropland. Many mechanisms are available on spreaders to distribute manure. The effectiveness of these mechanisms is dependent upon the characteristics of the manure being spread. The quality of the spreading is generally satisfactory for damp, uncompacted manure. Wet manure tends to drop in a windrow behind the spreader.

Open-tank spreaders generally are built with a shaft mounted near the open top and parallel to the main axis of the tank. Chains on this shaft act as flails when the shaft turns, and the manure is discharged out the side of the spreader.

Table 5
ESTIMATED ANNUAL MACHINE USAGE (HR/YEAR) FOR MANURE LOADING INTO BOX SPREADERS AND MANURE TRUCKS

Cattle feedlot capacity[a] (head)	Annual collectable manure quantity[b] (t/year)	Manure loading rate[c] (t/hr)						
		25	50	100	200	300	400	500
100	150	6	3	1.5	1	—	—	—
1,000	1,500	60	30	15	8	5	4	3
5,000	7,500	300	150	75	38	25	19	15
10,000	15,000	600	300	150	75	50	38	30
50,000	75,000	—	1,500	750	375	250	188	150
100,000	150,000	—	—	1,500	750	500	375	300

[a] Applies to beef feedlots or open lot dairy corrals; for sheep feedlots, 1 head cattle capacity = 6 sheep.
[b] Assumes collectable manure production of 5 kg/day (wet basis) and animal occupancy is 80% of feedlot capacity.
[c] Manure loading rate at actual field efficiency which may be as low as 25% of machine capability because of downtime between spreader vehicle arrival.

From Sweeten, J. M., Great Plains Beef Cattle Handbook L-1979, Tex. Agric. Ext. Service, Texas A & M University, College Station, 1979. With permission.

Table 6
TYPICAL SPECIFICATIONS FOR COMMERCIAL SOLID MANURE SPREADING EQUIPMENT

Specification item	Tractor-towed spreader		Truck-mounted spreader	
	Small	Large	Small	Large
Max. rated load capacity, t	4	10	10[a]	15[a]
New ASAE rating — struck level, m³	2.86	6.94	9.69	14.36
Inside width of bed, m	1.51	1.67	2.15	2.15
Overall width, m	2.35	2.77	2.42	2.42
Inside depth of bed, m	0.57	0.76	1.02[b]	1.02[b]
Inside length of bed, m	3.26	—	4.42	5.94
Overall length, m	4.95	7.50	5.02	6.55
Overall height, m	1.19	1.35	—	—
Wheel tread, m	2.04	2.34	—	—
Weight, kg	740	2030	—	—
No. beaters	1 or 2	1 or 2	2	2
Beater speed, rpm	—	360	400	400
Drive PTO speed, rpm	540 or 1000	540 or 1000	—	—
No. apron conveyor speeds	5	3	Infinite	Infinite
Recommended tractor size, kW	30—37	97	120[a]	—

[a] Estimated value.
[b] Without optional sideboards.

Box spreaders may be obtained as tractor-towed machines, or they may be mounted on trucks. Typical specifications for either type are shown in Table 6. Tractor-towed spreaders range in capacities from 2.5 to 18.5 m³, or 4 to 10 t.[1] The spreading mechanism on spreaders having capacities under 3.5 m³ is available with either ground-driven mechanisms or PTO drives (540 or 1000 rpm). PTO drives are used on spreaders having capacities over 3.9 m³.

Two or four wheels are used, depending on the capacity of the spreader and size of tires used.

Truck-mounted, box-type spreaders range in capacity from 6.2 to 18.5 m³. The smaller trucks are single axle while the larger trucks are tandem axle.

A combination of steel and wood is often used to fabricate spreader boxes. Paddles and tined augers are among the rear "beater" devices used to reduce particle size and distribute manure. Beater rotational speeds are 300 to 400 rpm. It is a common practice to provide for variable speed on the apron (0 to 4.4 m/min). Some spreaders have moving front-end gates. Features of modern box-type, tractor-towed manure spreaders include hydraulic end gates and front wall extensions to contain wet manure. Upper beater attachments can be added to enhance spreading of cohesive or hard-packed manure as well as manure containing a high proportion of straw.

Recommended tractor sizes for towed manure spreaders range from 30 kW for the smallest (4 t) spreaders to 97 kW for the largest (10 t) spreaders.

Time required for hauling and spreading solid manure varies according to distance, terrain, road quality, and equipment. Operator delays, refueling stops, loading time, and livestock interferences can cause wide variation between trips of equal distance. For both short- and long-haul distances, time is saved by using the largest manure spreader that is economically affordable, consistent with reasonably high annual use rate and prevention of soil compaction.

In most instances, manure is spread on cropland adjacent to or within 1 km of the animal feeding facility. In those instances, two to four loads per hour can be achieved.

Large commercial feeding operations often have insufficient land for manure diposal or utilization, in which case manure is hauled to cooperating farmers as far away as 40 km (one way). Where most of the travel is over all-weather roads, the mean round-trip cycle time required for hauling and spreading feedlot manure can be estimated from the following equation,[2] for distances ranging from 5 to 22 km:

$$T = 0.7 + 4.0D - 0.07D^2 \tag{2}$$

where T = round-trip cycle time, min, and D = one-way haul distance, km.

Energy requirements for feedlot manure transportation and spreading vary with travel distance and size of spreader truck. Fuel mileage for both single- and tandem-axle trucks is commonly poor (1.3 to 1.5 km/ℓ). Using a single-axle (9 t) spreader truck, manure transportation and spreading takes an estimated 14 and 50 kW-hr/t of manure for 8- and 32-km (one-way) haul distances, respectively. A tandem-axle 13-t truck requires 33% less energy and 30% lower cost per tonne at 8- and 32-km haul distances than the 9-t truck.[6] Calculated energy savings from use of 34-t semitrailer trucks for hauling farther than 8 km appear sufficient to offset the added energy requirements for on-farm reloading into spreader trucks or towed box spreaders.

REFERENCES

1. **Miner, J. R. and Smith, R. J., Eds.,** Livestock Waste Management with Pollution Control, North Central Regional Res. Publ. 222, Midwest Plan Service Handbook MWPS-19, Midwest Plan Service, Ames, Iowa, 1975, 23.
2. **Sweeten, J. M. and Reddell, D. L.,** Time-motion analysis of feedlot manure collection systems, *Trans. ASAE,* 22, 138, 1979.
3. **Sweeten, J. M.,** Manure Management for Cattle Feedlots, Great Plains Beef Cattle Handbook L-1769, Tex. Agric. Ext. Service, Texas, A & M University, College Station, 1979.

4. **Butchbaker, A. F., Garton, J. E., Mahoney, G. W. A., and Paine, M. D.,** Evaluation of Beef Cattle Feedlot Waste Management Alternatives, Rep. No. 13040FXG, Office of Research and Monitoring, U.S. Environmental Protection Agency, Washington, D.C., 1971, 122.

5. **Park, W. G., Jr.,** A Cost-Center Analysis for a Commercial Feedyard, Master's thesis, Texas A & M University, College Station, 1972.

6. **Sweeten, J. M., Reddell, D. L., and Stewart, B. R.,** Feedlot Manure as an Energy Source, paper presented at the 1974 Texas Section Meeting, American Society of Agricultural Engineers, Abilene, 1974.

DRYING OF WASTES

W. C. Fairbank

REASONS FOR DRYING

The acceptance of animal manure as a byproduct of beneficial use or its rejection as worthless waste and environmental insult is primarily related to its moisture content. Dry manure that is below 30% moisture, as-is, is not generally an unpleasant product to handle. Wet or moist manure generates odor, is a nuisance-fly substrate, and is aesthetically unacceptable. Free-liquid and cellular moisture are responsible for stickiness, poor physical properties, and excessive weight. Moisture dilutes the nutrient and energy values which might otherwise lead to profitable uses.

Feces and urine are usually deposited or voided onto the same manure collection surface, so both must be managed. For large animals, the mixed manure is about 88 to 90% water (10 to 12% total solids) before draining. Poultry excrete a mixed waste that averages about 80% moisture for healthy chickens (Figure 1). Wet stackable manure or fresh feces only from commercial livestock including cattle, sheep, swine, horses, and poultry is usually 70 to 80% moisture on an as-is basis.

Water from external sources such as rain, surface runoff, leaky drinking devices, and washing processes often falls onto manure collection areas or in other ways becomes mixed into the waste. This water is an energy debit that can make drying a rather impractical waste management scheme. It imposes its sensible heat and latent heat of vaporization on the drying process as an additional heat demand of about 2500 kJ/kg (1075 Btu/lb). All practical means to exclude water should be implemented. A manure accumulation that becomes unexpectedly saturated by rain or water should be eliminated from the process or set aside until it dewaters naturally to a manageable level. Manure wastes that are collected and transported as slurries or liquids should not be thought of as dryable because of very negative energy value.

DRYING PROCESS

Drying of manure uses the same heat transfer and mass flow principles and similar materials handling systems as the drying of forages or other relatively low-valued fibrous materials. Drying is an adiabatic process involving two steps that occur simultaneously: (1) vaporizaiton and (2) removal of water vapor. Step 1 is a transfer of sensible and latent heat energy into the wet material; step 2 is the convective dispersal of vapor into the surrounding atmosphere augmented by natural or forced air movement. Vaporization is a function of temperature and available heat to enter the mass through the mechanisms of conduction, convection, and radiation; dispersal is a function of vapor pressure (humidity) of the immediately surrounding air and through the boundary layer conditions as affected by velocity and turbulence.

Drying occurs naturally under the whims of nature whenever net evaporation exceeds precipitation. The energy and vapor transfers can be followed on a psychrometric chart, which relates heat (enthalpy), temperature, and air moisture content. Artificial drying (commonly called dehydration) amplifies the temperature and moisture differentials so that heat and vapor transfers are accelerated. It also provides a continuing input of thermal energy.

Effective drying is the management of temperature, heat, humidity, and air movement through and around the mass, and exhaust of the vapor-laden air. Natural drying exhibits strong diurnal and annual cycles as the forces are solar and weather related. Dehydration provides and maintains temperatures and air exchange rates at higher than ambient temper-

FIGURE 1.   Dehydrated poultry waste for cattle feed.

ature. Mechanical dehydrators include a product-conveying system so that the process is one of continuous flow. Most farm-type dehydrators use an exhaust temperature-modulated burner system.

Unheated forced-air dryers are returning to use as a simple way to conserve energy by substituting labor for purchased fuel. Apparatus to capture readily available solar heat in the system is being developed as another cost-effective dehydration scheme.

Final dryness must satisfy the purpose for drying, the specification for the end use of the product, or the preference of the customer if it is to be marketed. Overdrying is costly in terms of utilization of facilities and excessive energy consumption. It also reduces the weight of the dry product and thus loss of revenue and may lead to loss of use by the nursery trade, which desires a moist organic mulch.

Drying for control of nuisances while the product is in storage need only break the fly life cycle (flies lose interest below 35% moisture) and promote the formation of voids of aeration. This will prevent putrefaction and will also encourage aerobic composting — nature's best waste management scheme.

The feed ingredient market that is developing for dried animal waste protein is accustomed to buying dry commodities at a moisture content that is presumed to be close to the upper allowable limit for safe storage. For dried animal wastes that is near 10%. The desired final moisture for any byproduct uses should be specified by each market or user to preclude overdrying.

Market-grade manure for turf grass top dressing is usually used as a dry free-flowing product that can be spread by machine. For general garden use, it should be damp compost for spading into areas that are to be planted. The nursery trade has use for bulk manure for blending into planter mixes, and most nurseries retail bagged manure (or manure-based products) for home garden use. The bagged product is usually in storage at one or more sites between the producer and consumer. Manure that has not dried to dustiness or been composted to a biologically stable condition will have active microorganisms that release ammonia down to near 5% moisture. Damp manure rots paper bags rather quickly; therefore, most distribution is now in plastic bags or liners. The person in charge of distribution should determine the best dryness on the basis of customer desires, recognizing the cost of drying and the packaging requirements.

Thermal energy recovery from manure by direct combustion in large municipal incinerators or in commercial/residential-type burners is possible if the waste material is essentially dry so that it will maintain flame. The cellulosic portion of fresh dry cattle manure gives it a higher heating value, near 5000 Btu/lb of dry matter. Manure moisture content greater than 50% usually cools combustion below the ignition temperature so it will not burn. Manure can be net energy positive only if the original moisture is removed by natural drying. Cattle manure is high in undigested feed material and at 15% or less moisture can burn well and without odor in an air-regulated fire box. Swine and poultry manures, by comparison, are more highly mineralized and do not contain enough combustible matter to be considered practical fuels.

NATURAL DRYING

Natural drying occurs almost everywhere at times. In the semiarid sunbelt, it proceeds throughout the year with only temporary periods when precipitation exceeds evaporation. Natural drying of waste should be the first alternative treatment to be considered because it is the ultimate energy-conserving as well as nuisance-control process. Sunshine transfers heat into manure that is in the open surrounding atmosphere and takes up and disperses the released water vapor. Natural drying is optimized by selecting a site of maximum solaration and air movement. Trees or structures that cast shadows or restrict wind slow the drying process. An area that is open somewhat beyond the ultimate limits of the pile is needed. A slight knoll is best because it generally is subject to the highest available air velocity.

Thin-bed spreading (Figure 2) is a technique for natural drying in which the manure waste is spread each day in a layer up to 1-cm average thickness over an area allocated and properly prepared for drying. In dry climates, the moisture level diminishes within a day or two so that immature fly larvae cannot complete their growth phase. The combination of sunshine, drying, and air dilution usually precludes troublesome odors. A flail-type manure spreader is the only equipment needed for transport and spreading. Once-over-daily with a peg-tooth or spring-tooth harrow is usually desirable to break clumps and assure uniformity. As the manure reaches a semistable moisture content below 35%, it can be scraped into piles or windrows for final curing and storage. It composts well in this condition. The windrow should be tarped as needed for rain protection.

Tiller-drying (Figure 3) is an effective process for poultry manure under good drying conditions. The manure is dumped in small piles up to $1/_2$ m in height onto an area designated as the dry-yard. (The area is first smoothed and highly compacted to approach a floor in character.) Surface water intrusion must be precluded by grading, diversion ditch, or berm. A tractor tiller, spring-tooth harrow, ring harrow, or similar implement is used to spread the piles to a depth not greater than 10 cm. Daily or twice daily tilling is necessary to reaerate the mass and to expose wet material. Tilling also tends to macerate or desiccate fly larvae.

Tiller-drying machines have been constructed in the West and poultry producers in Japan now use them within greenhouse-like plastic rain shelters for year-round operation. Basically, steel, wood, or concrete tracks 3 m apart and up to 50 m in length confine a drying pad and support and guide a mechanized, carriage-suspended, motor-driven tiller. A control clock and reversing switches operate the machine on a cycle determined by environmental conditions and loading rate. Wet manure is dumped onto the pad at one end and dries as it automatically advances to the far end in about 3 to 10 days.

Windrowing manure during natural drying or curing periods provides the best conditions

FIGURE 2A.   Thin-beds spreading for rapid drying.

FIGURE 2B.   Daily harrowing speeds the drying and controls flies.

for encouraging composting and moisture loss. The irregular surface improves wind turbulence and aeration. Steep windrows will effectively shed limited rainfall.

Climate is the first criterion of natural drying. The cold, humid, Atlantic Northeast region in winter offers no hope of natural drying. The hot, arid Pacific Southwest in summer, where everything dries quickly if nothing is done, is the opposite situation. All regions and all seasons are somewhere between these environmental extremes. A natural drying system for part of the year, backed up by a fueled system for use during poor drying periods, can usually be cost effective. Net positive pan evaporation as measured by a National Weather Service, Class-A Evaporation Pan usually means that natural drying is potentially possible. The local Weather Service office can provide evaporation data.

Solar intensification onto manure drying pads by the use of adjacent reflective or radiating surfaces has been used for drying small lots. This arrangement is a passive solar system.

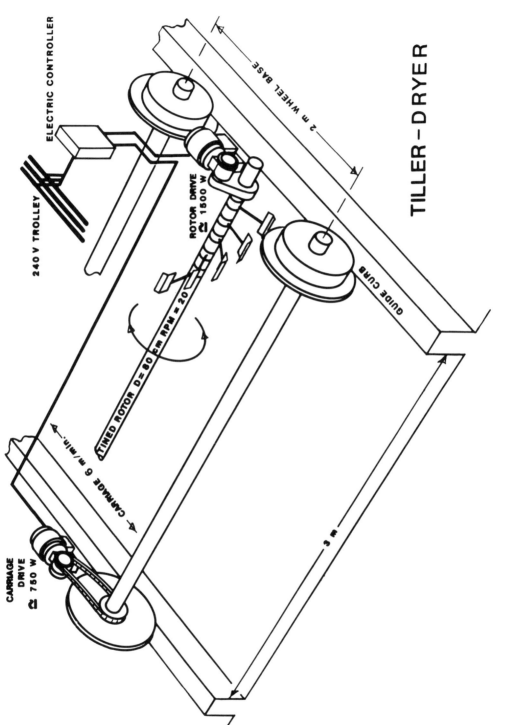

FIGURE 3.   The tiller-dryer concept.

Unfortunately, it is difficult to keep vertical walls or elevated reflecting surfaces from interfering with air movement. Investigations are underway to perfect the use of solar-heated air for manure drying. This would be an active solar system.

ARTIFICIAL DRYING

Artificial drying is the application of heat for vaporization of the embodied moisture and a system, usually forced air, to remove the water vapor. It usually also includes a mechanized system to move wet material into and dry product away from the dehydrator. Manure has been dried experimentally in probably every kind of dehydrator that has the ability to handle its difficult adhesive, cohesive, abrasive, corrosive properties and its general fibrous character and heterogeneity. Many unusual industrial-type dehydrators including ram-jets, under-fired conductive dryers, overhead IR tubes, freeze-vac driers, vacuum spray chambers, and microwave tunnels have been demonstrated, but with generally negative practicality. The value in dried manure has not been able to justify the high cost of advanced technology industrial dryers.

Existing farm drying equipment including forage dryers, gravity flow grain dryers, fruit tray dehydrator tunnels, and pot-hole bin dryers have been tried or have seen limited use during the normal commodity off-season. Drying performance was generally satisfactory, but labor efficiency was poor, equipment was severely abused, and sanitation code violations were suspected.

Rotary drum dryers have proved most successful for manure dehydration under all situations, given the product handling properties and the equipment capital and direct costs. Direct-fired, or single-pass dryers are now most common. They can be set up to burn oil or natural or LP gas with few changes of parts. Their heavy steel construction generally resists the abrasion of manure, high fire box temperature at one end and corrosive condensate at the other end. Used rotary drum dryers acquired from other industries have been the primary factor of economic success of many manure dehydrating plants.

Parallel-flow operation where the product and the hot gas move in the same direction allows high feed-end temperature and maximizes throughput capacity. Wet manure can be introduced very close to the flame because wet material will not burn. Partially dried manure or manure containing dry litter must be introduced further from the flame to prevent charring and the unpleasant odor of burned hair or feathers. Concentric drum and multipass dryers, although of higher thermal efficiency, usually have excessive maintenance costs.

Commercial dryers used for manure are usually about 2 to 3 m in diameter, 8 to 18 m long, single-pass design, and have fuel rates up to 10×10^6 kJ/hr. The product leaves the drum at about 60 to 80°C. An air conveyer system removes additional moisture and drops the temperature as the product is conveyed towards storage bins. The fines are removed by cyclone and are blended into the wet manure feed. This improves the condition for dehydration by distributing the moisture through greater bulk.

Stack emissions from a manure dryer require close attention for odors. Because few dehydrator operations can be atmospherically isolated from residential communities, the possibility of neighborhood complaint is always present. The plume is mostly water vapor, which precipitates as a mist in humid wather or quickly evaporates in dry atmospheres. However, there is often some blue smoke from incineration or volatilization or organic matter with its unpleasant odor. Odor control chemicals (perfumes, reactants, maskants, reodorants, etc.) have not generally been a satisfactory solution. Those that are reasonably effective become a burdensome expense. Artificial fragrances usually become almost the same as the original odors they are to mask or neutralize.

Flue gas scrubbing (Figure 4) to remove all pollutants may be specified by the local air quality agency. Condensation of the water vapor plume in order to deal with the noncon-

FIGURE 4.   A flue-gas scrubber on a manure dehydrator.

densibles is the first step. The effectiveness of a cold water spray for this purpose can be predetermined from psychrometric conditions. Dissipation of the heat of condensation absorbed in the water is often difficult and can create a new problem with biological dimensions. Recirculation through a wastewater pond will provide an effective heat sink and will also settle out trapped dust. Unfortunately, the organic burden will eventually cause heavy microbial growth and sludging unless the water is frequently spread over a grassland filter or used for irrigation. The emissions from noncondensibles can usually be scrubbed adequately with an alkaline solution. A simple do-it-yourself scrubber uses tightly stacked bottomless plastic milk bottles in a vertical containment vessel to give a large area of wetted gas absorption surface and unrestricted air flow. The odorous gas flows upward and out, while the solution is sprayed onto the top tier to trickle downward. (The effluent is nutrient entriched and requires proper management.)

SUMMARY

All livestock operations need a wet-season liquid waste handling system, and most can justify a dry-season ratio. The duration and intensity of wet weather are the primary determinants of the overall management plan as related to dryability to waste or to its ultimate utilization or disposal. Low rainfall regions or climates that have dependable dry seasons should utilize the drying forces of nature's sun, wind, unsaturated atmosphere, and aerobic microorganisms. Artificial drying is energy intensive and should only be used as backup to natural drying. Natural and artificial drying technologies can usually be combined to give greater operational and energy efficiencies than either alone. The total energy value represented in dry manure at its final point of use as compared to the caloric and embodied energy in substitute products and their handling methods should be the criteria for selecting and managing a manure drying operation.

ODORS AND NUISANCES

J. R. Miner

ODORS AND LIVESTOCK PRODUCTION

The design and management of livestock production facilities have a major impact on the intensity and quality of odor released. With mounting pressures for a clean environment, the coincidental desire for freedom from imposition of neighbor's influence, and the trend to confine livestock production, the need for odor control technology has assumed an increasingly important status. The need for odor control technology is reflected in the frequency of odor-based complaints to air pollution control agencies, to individual producers, and, in the extreme, official damage complaints and suits leading to court action.

Livestock production odors, like other nontoxic odor emissions, are generally regarded as nuisance pollutants. As such, they are not currently regulated by federal action under the Clean Air Act, but are the subject of an increasing number of state and local regulations and ordinances. Such legal processes may specifically regulate odors by definition of acceptable intensity limitations, durations, and frequencies to constitute a nuisance at the property line, or odor problems may be indirectly controlled by restricting the size, design, or location of livestock production enterprises. The latter approach is generally part of an overall land use plan and may not be recognized as specific odor regulation.

The vast majority of livestock and poultry production enterprises in the U.S. operate without experiencing odor complaints or causing concern among their neighbors. Odor conflicts are most frequent among new, large or recently expanded operations located near housing areas, shopping facilities, or other points of high sensitivity. It is also noteworthy that odor conflicts are generally accompanied by other signs of neighbor or community dissatisfaction. Accompanying complaints generally include water pollution unsightliness, noise, or insect nuisances.

Odorous compounds evolved from livestock production have never been measured in physiologically toxic concentrations and are not generally considered hazardous to human health in the concentrations normally encountered within a livestock building or adjacent to an outdoor feedlot. Under certain situations (such as manure pit agitation), however, dangerous gas concentrations can develop. Livestock odors, therefore, are nuisance pollutants and, like other nonhazardous assaults to the environment, must be regarded accordingly. Of principal importance are intensity, duration, and frequency of perception. More than 40 odorous compounds have been identified in the air near manure storage or treatment devices (Tables 1 and 2). Many of these compounds are known to be odorous in trace concentrations (Table 3). The maximum concentrations at which some of these compounds are considered nonhazardous to human health are shown in Table 4.

Most quantitative measurements of odorant concentrations suggest that the odor perceived from livestock production enterprises is a result of the mixture of odorous compounds, since all of the measured compounds are present in concentrations below their threshold. Table 5 includes odorous gas concentration measurements under a variety of circumstances.

RELEASE RATES AND MEASUREMENT

Documentation of specific odorant release rates has not been accomplished relative to the livestock industry to the same extent as from other sources. Table 6 summarizes ammonia evolution rates that have been measured under a variety of livestock production situations.

Table 1
COMPOUNDS IDENTIFIED IN THE AIR FROM THE
ANAEROBIC DECOMPOSITION OF LIVESTOCK AND
POULTRY MANURE

Alcohols	Acids	Amines	Fixed gases
Methanol	Butyric	Methylamine	Carbon dioxide
Ethanol	Acetic	Ethylamine	Methane
2-Propanol	Propionic	Trimethylamine	Ammonia
n-Propanol	iso-Butyric	Triethylamine	Hydrogen sulfide
n-Butanol	iso-Valeric		
iso-Butanol			
iso-Pentanol			

Carbonyls	Esters	Sulfides
Acetaldehyde	Methyl formate	Dimethyl sulfide
Proponaldehyde	Methyl acetate	Diethyl sulfide
Butyraldehyde	iso-Propyl acetate	Disulfides
iso-Butyraldehyde	iso-Butyl acetate	Mercaptans
Hexanal	iso-Propyl propionate	Methyl mercaptan
Acetone	Propyl acetate	Nitrogen heterocycles
3-Pentanone	n-Butyl acetate	Indole
Formaldehyde	p-Cresol	Skatole
Heptaldehyde		Pyrazines
Valeraldehyde		
Octaldehyde		
Decaldehyde		
Diacetyl (2,3-diketo-butane)		

From Miner, J. R. and Smith, R. J., Midwest Plan Handbook MWPS-19, Midwest Plan Service, Ames, Iowa, 1975. With permission.

They are characterized by extensive variability, which further reflects the conditions under which the phenomenon occurs.

Various schemes have been proposed for the measurement of odors (Table 7). The concentration of the individually odorous compounds has been used by some investigators as an indication of odor transport and odor pollution. Ammonia and hydrogen sulfide have been widely used for this application because of their ease of measurement and known odorous characteristics. In the concentrations measured, however, they would not be detectable by the human nose if they were not accompanied by other odorous compounds. Studies by Hill and Barth[1] indicate the synergistic nature of the combination of the common manure odors.

The measurement of odor intensity based upon the number of dilutions required to reduce the concentration to a barely detectable level has been the most generally accepted method for evaluating odor concentrations. The Scentometer as distributed by Barneby-Cheney and described by Rowe[2] has been used by researchers[3] and, more recently, regulatory agencies[4] to evaluate odor control techniques and odor transport. The Scentometer is a device for dilution of odorous air with odor-free air to determine the number of dilutions necessary to reach the barely detectable level. Similar measurements have also been made in the laboratory using more sophisticated devices[5] and in a mobile unit described by Lidvall et al.[6] The use of cotton fabric swatches exposed in air has been used by Licht and Miner[7] as a technique to sample odor intensity and quality.

Table 2
VOLATILE COMPOUNDS IDENTIFIED IN
THE AIR OF SWINE CONFINEMENT UNITS
AND IN ANAEROBICALLY STORED
PIGGERY WASTES

Component	Ref. — air[a]	Ref. — waste[a]
Methanol	15	19
Ethanol	15	19
1-Propanol	15	19
2-Propanol	15	19
1-Butanol	15, 16	19
2-Butanol	16	19
2-Methyl-1-propanol	15	19
3-Methyl-1-butanol	16	19
2-Ethoxy-1-propanol	16	
2-Methyl-2-pentanol	14	
2,3-Butanediol	16	
3-Hydroxy-2-butanone	16	19
Propanone	16	
2-Butanone	14	
3-Pentanone	14, 16	19
Cyclopentanone	14	
2-Octanone	14, 16	19
2,3-Butanedione	14, 16, 17, 18	
Methanal	15	
Ethanal	15, 16	
Propanal	15, 16	
Butanal	15	
Pentanal	15	19
Hexanal	15	19
Heptanal	15	19
Octanal	15	19
Decanal	15	19
2-Methyl-1-propanal	15	
Methanoic acid	*14*	19
Ethanoic acid	16, 18	*19*
Propanoic acid	*14*, 16, 18	*19*
Butanoic acid	14, 16, 17, *18*	*19*
2-Methylpropanoic acid	14, 16, 18	*19*
Pentanoic acid	14, 16, 18	*19*
2-Methylbutanoic acid		*19*
3-Methylbutanoic acid	18	*19*
Hexanoic acid	18	19
2-Methylpentanoic acid		*19*
4-Methylpentanoic acid	18	19
Heptanoic acid	16, 18	19
Octanoic acid	18	19
Nonanoic acid	18	19
Ethylformate		19
Methylacetate		19
Ethylacetate	14	19
Propylacetate		
Butylacetate		19
i-Propylacetate		19
i-Butylacetate		19
i-Propylpropionate		

Table 2 (continued)
VOLATILE COMPOUNDS IDENTIFIED IN THE AIR OF SWINE CONFINEMENT UNITS AND IN ANAEROBICALLY STORED PIGGERY WASTES

Component	Ref. — air[a]	Ref. — waste[a]
Phenol	*14*, 16, *18*	*19*, 24
3-Methylphenol		*19*, 24
4-Methylphenol	*14*, 16, 17, *18*	*19*, 24
4-Ethylphenol	*14*, 16	*19*
Toluene	14, 16	19
Xylene	14, 16	
Indane	16	
Benzaldehyde	14, 16	19
Benzoic acid	16	*19*, 24
Methylphtalene	16	19
Indole	*18*	19
Skatole	*14*, 18	*19*
Acetophenone	14, 16	19
Phenylacetic acid		17, *19*, 24
3-Phenylpropionic acid		
		17, 19, 24
Ammonia	16	
Methylamine	20	
Ethylamine	20	
Trimethylamine	16	
Triethylamine	20	
Carbonylsulfide	21	
Hydrogen sulfide	18, 21, 22	19, 23
Methanethiol	21	23
Dimethylsulfide	21	19
Diethylsulfide		19
Dimethyldisulfide	16, 21	19
Dimethyltrisulfide	16	19
Ethanethiol		19, 23
Diethyldisulfide	*14*	19
Propanethiol	*14*	23
Butanethiol	*14*	23
Dipropyldisulfide	*14*	
2-Methylthiophene	*14*	
Propylprop-1-enyldisulfide	*14*	
2,4-Dimethylthiophene	*14*	
2-Methylfuran	*14*	

[a]　Components with references in italics are considered by the respective author as main constituents responsible for the offensive odors from piggery wastes.

From Spoelstra, S. F., Microbial Aspects of the Formation of Malodorous Compounds in Anaerobically Stored Piggery Wastes, Agricultural University, Wageningen, The Netherlands, 1979. With permission.

Table 3
CHARACTERISTICS AND PERCEPTIBLE
CONCENTRATIONS OF VARIOUS
SUBSTANCES IN AIR[24]

Substance	Odor characteristic	Concentration causing faint odor (10^{-9} g/ℓ)
Acetaldehyde	Pungent	4
Ammonia	Sharp, pungent	37
n-Butyl mercaptan	Strong, unpleasant	1.4
Carbon disulfide	Aromatic odor, slightly pungent	2.6
Ethyl mercaptan	Odor of decayed cabbage	0.19
Hydrogen sulfide	Odor of rotten eggs, nauseating	1.1
Methyl mercaptan	Odor of decayed cabbage or onions	1.1
Propionaldehyde	Acrid, irritating odor	2
Propyl mercaptan	Unpleasant odor	0.075

Table 4
THRESHOLD LIMIT VALUES
FOR VARIOUS GASES
ASSOCIATED WITH ANIMAL
WASTE ODORS[25]

Substance	Concentration in air (10^{-9} g/ℓ)
Acetaldehyde	360
Acetic acid	25
Ammonia	35
n-Butyl acetate	710
Butyl mercaptan	35
Diethylamine	75
Dimethylamine	18
Ethylamine	18
Ethyl mercaptan	25
Isopropylamine	12
Methylamine	12
Methyl mercaptan	20
Triethylamine	100

Note: The Threshold Limit Values refer to air-borne concentrations under which it is believed that nearly all workers may be repeatedly exposed without adverse effect.

Table 5
CONCENTRATIONS OF ODOROUS COMPOUNDS MEASURED IN THE VICINITY OF LIVESTOCK PRODUCTION FACILITIES

	Constituents		
Sample source	NH_3 (mg/m³)	Total sulfides (mg/m³)	Ref.
Broiler production room litter system	2.38		26
Texas high plain beef feedlot	0.12	0.005 and 0.027	27
Swine confinement building	4—24		20
Poultry house	0.45—2.6	1.0—14.9[a]	28
Swine confinement building	7.4	0.10	29
Swine confinement building		0.12—0.85[a]	38

[a]   Hydrogen sulfide concentration in parts per million by volume.

Table 6
AMMONIA VOLATILIZATION FROM VARIOUS SURFACES ASSOCIATED WITH LIVESTOCK PRODUCTION

Situation	Rate (kg/ha/day)	Ref.
Manure-free pasture (Oregon)	0.01—0.02	30
Pasture (Nebraska)	0.044	31
Pasture after recent manure application	0.05—0.2	30
Sheep pasture, late summer	0.26	32
Beef feedlot (Nebraska)	0.41	31
Manure covered aisle in dairy barn	0.5—1.0	30
Beef feedlot (Idaho)	3.48 average 0.31—12.2 range	33
Anaerobic swine manure lagoon	17—98	34

ODOR CONTROL PRINCIPLES

Techniques for the control of odor release are based upon a limited number of rather specific control principles. Some volatile compounds present in the feed, manure, or manure slurry can be converted to a less volatile form by pH control, chemical conversion, or biological conversion to a less odorous or less volatile compound. Examples of this approach would include the addition of lime as a pH-adjusting material to control the release of hydrogen sulfide.[8] The dissociation of hydrogen sulfide is a strong function of pH, and if the pH is raised above 9.5, the escape of hydrogen sulfide should be insignificant. Para-formaldehyde has been added to manure as a means of converting ammonia to nonvolatile

Table 7
TECHNIQUES THAT HAVE
BEEN USED TO MEASURE THE
CONCENTRATION OF
AIRBORNE CONSTITUENTS OF
LIVESTOCK WASTE ODORS

Constituent	Technique[a]	Ref.
Particulate matter	a	
	b	
Ammonia	c	
	d	20
	e	9
	f	
Hydrogen sulfide	c	
	d	
	e	
	f	
Odor intensity	Scentometer	
Carbonyls	g	16
Alcohols	g	15

[a] a = commercially available direct reading instrument;
   b = filtration and weighing;
   c = colorimetric indicator tubes;
   d = selective absorption preceding analysis of the liquid;
   e = indicator papers which develop color in response to the constituent of concern;
   f = chromatographic procedures;
   g = selective absorption preceding chromatographic separation.

hexamethylenetetramine.[9] Sulfides are oxidized to sulfates under aerobic waste treatment, as is practiced in oxidation ditches and aerated lagoons.

Another approach to the control of odor is to inhibit the anaerobic decomposition of manure. This is most frequently accomplished by keeping feedlots sufficiently dry to allow oxygen permeation of the surface. In systems utilizing bedding, the bedding serves as a similar role. In liquid systems, odorant-producing decomposition is most often inhibited by maintaining the slurry in an aerobic condition. Oxidation ditches and aerated lagoons are two examples of this practice. Bacterial decomposition of manure is also inhibited by low temperature, but, except in site selection, little use can be made of this fact.

Physical confinement of the odorants offers a third potential for odor control. Covers on manure storage tanks and anaerobic manure treatment devices are effective in controlling the escape of odorants. By reducing the air exchange over manure slurries, the volatilization of odorous gases is reduced. Gases escaping from enclosed tanks may be further managed by incineration, liquid scrubbing, or soil column absorption. Although no entirely satisfactory low-cost cover material has been developed, covering of anaerobic lagoons and venting the released gases to a burner or a soil absorption field offers potential for odor control.

Another option for reducing odor complaints associated with confinement livestock operations is to reduce the odor of exhaust air. Hammond et al.[10] have indicated that swine confinement building odors are associated with airborne particles. Willson[11] installed a dust removal unit on a poultry house exhaust fan and achieved a particle concentration reduction;

Table 8
SULFUR-CONTAINING VOLATILES IN ANIMAL MANURE VOLATILES AND THEIR POSSIBLE PRECURSORS

Products	Possible precursors
H_2S	Cysteine, cystine (35), sulfate
Methylmercaptan	Methionine, methioninesulfoxide, methioninesulfone, S-methylcysteine (21)
Ethylmercaptan	Ethionine, S-ethylcysteine (21)
Dimethyl sulfide	Methionine, methioninesulfoxide, methioninesulfone, S-methylcysteine, homocysteine (21)
Dimethyl disulfide	Methionine, methioninesulfoxide, methioninesulfone, S-methylcysteine (21)
Diethyl disulfide	Ethionine, S-ethylcysteine (21)
Methylethyl sulfide	Ethionine, S-ethylcysteine (21)
Carbon disulfide	Cysteine, cystine, lanthionine, djenkolic acid, homocysteine (21)
Carbon sulfide	Lanthionine, djenkolic acid (21)
Methylthioacetate	Methionine + glucose (36)
Diethyl sulfide	Cysteine, cystine
Dipropyl disulfide	Propylcysteine
Diallyl disulfide	Allycysteine

From Spoelstra, S. F., Microbial Aspects of the Formation of Malodorous Compounds in Anaerobically Stored Piggery Wastes, Agricultural University, Wageningen, The Netherlands, 1979. With permission.

Table 9
EFFECTS OF VARIOUS MANURE VOLATILES ON HUMANS

	Concentration (ppm)	Exposure period (min)	Physiological effect
Ammonia			Irritant
	400		Irritation of throat
	700		Irritation of eyes
	1,700		Coughing and frothing
	3,000	30	Asphyxiation
	5,000	40	Could be fatal
Carbon dioxide			Asphyxiant
	20,000		Safe
	30,000		Increased breathing
	40,000		Drowsiness, headaches
	60,000	30	Heavy, asphyxiating breathing
	300,000	30	Could be fatal
Hydrogen sulfide			Poison
	100		Irritation of eyes and nose
	200	60	Headaches, dizziness
	500	30	Nausea, excitement, insomnia
	1,000		Unconsciousness, death
Methane			Asphyxiant
	500,000		Headaches, nontoxic

From Francis, A. J., Duxbury, J. M., and Alexander, M., *Soil Biol. Biochem.*, 7, 51, 1975. With permission.

Table 10
SUMMARY OF ODOR INTENSITIES AT 17 TEXAS
CATTLE FEEDLOTS, 1973—75

Location	Condition	Odor intensity, DT
Feedlot surface	Dry	7
	Well drained, moist	7—31
	Poorly drained, damp	31
	Wet, ponded	31—170
Runoff retention ponds	—	31—170
Manure stockpiles	—	7—170

From Avery, G. L., Merva, G. E., and Gerrish, J. B., *Trans. ASAE,* 20, 3, 1977.
With permission.

however, the device was not effective in odor reduction. Licht and Miner[7] built and tested a cross-current wet-packed bed scrubber which effectively reduced both particle concentrations and odor intensity. The unit they devised is similar in concept to one previously described by van Geelan and van der Hoek[12] in the Netherlands.

DESIGN AND MANAGEMENT PRACTICES TO CONTROL ODORS

Application of odor control techniques requires specific attention to the operation under discussion. Perhaps the most critical and effective means of reducing odor complaints occurs in the initial site selection. Although it is difficult to establish definitive perimeters beyond which odor complaints will not be problems, a livestock producer must seriously consider odor control as he selects a site. A site may be ideally suited for livestock production in terms of transporation, feed supply, and zoning regulations, but may be inappropriate because of existing or proposed development in the area.

Although wind direction is important in evaluating a site in terms of odors, most locations have winds from several directions during the year. The simple location "downwind" of development is not sufficient to assure acceptability. Where distance alone is used as the criterion, it must be expected that under appropriate climatic conditions, odors can be transported in excess of a mile downwind. If these conditions are sufficiently rare and the insult is slight, this may not be an inhibiting factor toward development.

The second opportunity for reducing odor problems occurs during the design and construction of a facility. By application of odor control principles, the probability of odor production can be minimized. Designing outdoor lots that are well drained, watering systems that do not flow onto the lot surface, and runoff control facilities that are remotely located from areas of odor sensitivity will achieve some odor reduction. In confinement facilities, the methods of manure removal from the pens, manure transport, and the handling approach are most important for odor control. Also, the animals must be kept clean. Among approaches used for accomplishing this are slotted floors, flushing gutters, and frequent pen scraping. Covered manure storage tanks control odor release from stored manure. Where treatment is required and odor control is important, aerobic systems such as oxidation ditches and floating surface aerators, although more expensive, can be effectively used to maintain low odor intensities.

The operation and management of a livestock production facility also offers considerable opportunity for exercising odor control. Maintaining the operating system in functional order is probably the most important factor. Overflowing manure storage tanks, broken scrapers, leaking waterers, and ruptured retention ponds and dikes are among the most common causes of odor complaints.

Anaerobic lagoons for swine waste treatment are of special concern in odor control. Properly designed and managed lagoons are not free of odors but are infrequently the cause of serious odor problems. However, overloaded or shock-loaded lagoons are more likely to have objectionable odors. Where multiple-celled lagoons are used, it is important that the cell or cells receiving fresh manure not be loaded in excess of the recommendations for a particular area. Anaerobic lagoon odors are most common in the late spring and early summer when the water temperature warms and manure accumulated during the winter undergoes rapid decomposition. Where odor control is critical, it has been found helpful to remove supernatant and refill to the normal operating level with clean water. Another common recommendation where odor control is critical is to increase the normal design volume of the lagoon and thereby reduce the loading rate.

Where practical, it is desirable to locate lagoons as far as possible from neighboring residences, roads, and other odor-sensitive areas. Minimum recommendations range from 300 ft to as much as $1/2$ mi. Shielding lagoons from view is also helpful.

Manure disposal techniques and timing are also very important for odor control. When manure is to be applied to cropland, selection of a field downwind of neighboring residences on the particular day is important. Morning application of manure is more desirable than late afternoon application, which limits potential drying time. Neighbors are generally most sensitive to odor problems in early evening when utilizing outdoor recreational facilities.

When manure disposal is necessary and odor control is critical, immediate covering of the manure with soil can effectively minimize odor complaints. Where soil is suitable and neighbors are particularly close, direct manure injection beneath the soil surface is a valuable technique.

REFERENCES

1. **Hill, D. T. and Barth, C. L.,** Quantitative prediction of odor intensity, *Trans. ASAE,* 19(5), 939, 1976.
2. **Rowe, N. R.,** Odor control with activated charcoal, *J. Air Pollut. Control Assoc.,* 13, 150, 1963.
3. **Sweeten, J. M., Reddel, D. L., Schake, L. M., and Garner, B.,** Odor Intensity at Cattle Feedlots, presented at the 1st Annu. Symp. Air Pollution Control in the Southwest, Texas A & M University, College Station, November 5 to 7, 1973.
4. **Sweeten, J. M.,** Environmental Protection Requirements for Swine Operations, Fact Sheet L-1302, Texas Agric. Ext. Serv., 1974.
5. **Moncrieff, R. W.,** An instrument for measuring and classifying odors, *J. Appl. Physiol.,* 16, 742, 1961.
6. **Lindvall, T., Noren, O., and Thyselius, L.,** Odor reduction for liquid manure systems, *Trans. ASAE,* 17(3), 508, 1974.
7. **Light, L. A. and Miner, J. R.,** A Scrubber to Reduce Livestock Confinement Building Odors, ASAE Paper No. PN 78-203, American Society of Agricultural Engineers, St. Joseph, Mich., 1978.
8. **Day, D. L.,** Liquid hog manure can be deodorized by treatment with chlorine or lime, *Ill. Res.,* 127 (summer), 16, 1966.
9. **Seltzer, W., Moum, S. G., and Goldhaft, T. M.,** A method for the treatment of animal wastes to control ammonia and other odors, *Poult. Sci.,* 48, 1912, 1969.
10. **Hammond, E. G., Fedler, C., and Jung, G. A.,** Identification of Dust-Borne Odors in Swine Confinement Facilities, ASAE Paper No. 77-3550, American Society of Agricultural Engineers, St. Joseph, Mich., 1977.
11. **Willson, G. B.,** Control of odors from poulrty houses, in Livetock Waste Management and Pollution Abatement, ASAE Publ. PROC-271, American Society of Agricultural Engineers, St. Joseph, Mich., 1971, 114.
12. **van Geelan, M. A. and van der Hoek, K. W.,** Odor control with biological washers, *Agric. Environ.,* 3(2, 3), 217, 1977.
13. **Miner, J. R. and Smith, R. J.,** Livestock Waste Management with Pollution Control, Midwest Plan Service Handbook MWPS-19, Midwest Plan Service, Ames, Iowa, 1975.
14. **Spoelstra, S. F.,** Microbial Aspects of the Formation of Malodorous Compounds in Anaerobically Stored Piggery Wastes, Agricultural University, Wageningen, The Netherlands, 1979.

15. **Merkel, J. A., Hazen, T. E., and Miner, J. R.,** Identification of gases in a swine confinement building atmosphere, *Trans. ASAE,* 12, 310, 1969.
16. **Hartung, L. D., Hammond, E. G., and Miner, J. R.,** Identification of carbonyl compounds in a swine building atmosphere. Livestock waste management and pollution abatement, Proc. Int. Symp. Livestock Wastes, American Society of Agricultural Engineers, St. Joseph, Mich., 1971, 105.
17. **Hammond, E. G., Jung, G. A., Kuczala, P., and Kozel, J.,** Constituents of swine house odors, in Proc. Int. Symp. Livestock Environment, 1974, 364.
18. **Schaefer, J., Bemelmans, J. M. H., and Ten Noever de Brauw, M. C.,** Onderzoek naar de voor de stank van varkensmesterijen verantwoordelijke componenten, *Landbouwkd. Tijdschr.,* 86, 228, 1974.
19. **Schreier, P.,** Gaschromatographisch-massenspektrometrische Untersuchungen von Geruchstoffen aus der Tierhaltung, *VDI Ber.,* 226, 127, 1975.
20. **Miner, J. R. and Hazen, T. E.,** Ammonia and amines. Components of swine-building odor, *Trans. ASAE,* 12, 772, 1969.
21. **Banwart, W. L. and Bremner, J. M.,** Identification of sulfur gases evolved from animal manures, *J. Environ. Qual.,* 4, 363, 1975.
22. **Day, D. L., Hansen, E. L., and Anderson, S.,** Gases and odors in confinement swine buildings, *Trans. ASAE,* 8, 118, 1965.
23. **Janowski, T., Gawlik, J., and Zimnal, S.,** Gas-chromatographische Untersuchungen der Stalluft, *VDI Ber.,* 226, 123, 1975.
24. **Yasuhara, A. and Fuwa, K.,** Odor and volatile compounds in liquid swine manure. I. Carboxylic acids and phenols, *Bull. Chem. Soc. Jpn.,* 50, 731, 1977.
25. **Sheehy, J. R., Achinger, W. C., and Simon, R. A.,** Handbook of Air Pollution, Environmental Health Series, Air Pollution PHS Publ. No. 999-AP-44, Public Health Service, undated, A-17.
26. Threshold Limit Values for 1967, American Conference of Governmental Industrial Hygienists.
27. **Koelliker, J. K., Miner, J. R., Hellickson, M. L., and Nakaus, H. S.,** A Ziolite Packed Scrubber to Improve Poultry House Environments, ASAE Paper No. 78-4044, presented at the 1978 Meeting of the American Society of Agricultural Engineers, Logan, Utah, 1978.
28. **Peters, J. A. and Blackwood, T. R.,** Source Assessment Beef Cattle Feedlots, Environ. Protect. Technol. Ser. EPA-600/2-77-107, U.S. Environmental Protection Agency, Washington, D.C., 1977.
29. **Burnett, W. E.,** Air pollution from animal wastes. Determination of malodors by gas chromatographic and organoleptic techniques, *Environ. Sci. Technol.,* 3, 744, 1969.
30. **Lebeda, D. L. and Day, D. L.,** Waste caused air pollutants are measured in swine buildings, *Ill. Res.,* p. 15, Fall 1965.
31. **Miner, J. R., Kelly, M. D., and Anderson, A. W.,** Identification and measurement of volatile compounds within a swine confinement building and measurement of ammonia evolution rates from manure covered surfaces, in Managing Livestock Wastes, Proc. 3rd Int. Symp. Livestock Wastes, ASAE Publ. PROC-275, American Society of Agricultural Engineers, St. Joseph, Mich., 1975, 351.
32. **Eliott, L. F., Schuman, G. E., and Viets, F. G., Jr.,** Volatilization of nitrogen containing compounds from beef cattle areas, *Soil Sci. Am. Proc.,* 35, 752, 1971.
33. **Denmead, O. T., Simpson, J. R., and Freney, J. R.,** Ammonia flux into the atmosphere from a grazed pasture, *Science,* 185, 609, 1974.
34. **Miner, J. R. and Stroh, R. C.,** Controlling feedlot surface odor emission rates by application of commercial products, *Trans. ASAE,* 19, 533, 1976.
35. **Koelliker, J. K. and Miner, J. R.,** Desorption of ammonia from anaerobic lagoons, *Trans. ASAE,* 16(1), 148, 1973.
36. **Freney, J. R.,** Sulfur-containing organics, in *Soil Biochemistry,* McLaren, A. D. and Petersen, G. H., Eds., Marcel Dekker, New York, 1967, 229.
37. **Francis, A. J., Duxbury, J. M., and Alexander, M.,** Formation of volatile organic products in soils under anaerobiosis. II. Metabolism of amino acids, *Soil Biol. Biochem.,* 7, 51, 1975.
38. **Taigonides, E. P. and White, R. K.,** The meance of noxious gases in animal units, *Trans. ASAE,* 12(3), 359, 1969.
39. **Avery, G. L., Merva, G. E., and Gerrish, J. B.,** Hydrogen sulfide production in swine confinement buildings, *Trans. ASAE,* 18(1), 149, 1975.
40. **Sweeten, J. M., Reddell, D. L., Schake, L., and Garner, B.,** Odor intensities at cattle feedlots, *Trans. ASAE,* 20, 3, 1977.

BYPRODUCT RECOVERY

Philip R. Goodrich

The recovery of byproducts from livestock wastes has traditionally included any beneficial use of the livestock waste. Several byproducts and the processes associated with their recovery are so important that separate sections are included in this handbook for each byproduct.

Livestock wastes are a mixture of digested feed, spilled feed, bedding, secretions, microorganisms, and various inert materials. This complex mass of material is extremely variable in composition as produced by different species under various housing situations.[1] The list of useful products that have been isolated from livestock wastes is very long.[2,3] However, many of the byproducts that have been identified exist either in small quantities or are of low value.[4] Therefore they cannot be economically reclaimed. Examples are vitamins B-12 and B-2, gases such as hydrogen sulfide and ammonia, and minerals such as sodium. Other byproducts do exist in more usable forms and quantities.

Energy for heating is an available byproduct of livestock wastes. The organic material in the waste can be converted to heat by incineration (direct burning) if the moisture content of the waste is low enough. The net energy released by incineration is the total energy released minus the energy required to vaporize the water in the waste. Therefore, very wet waste is often unable to sustain combustion.[5] Composed livestock waste can be used as a fuel. The composting releases much of the water in the livestock waste and the resultant material will burn more readily. No large-scale experiments have been carried out, and this is not being commercially applied at the present time.

The incineration of animal waste to gain heat is a very old process. Ancient civilizations burned cattle dung for heat. It is used today by millions of people in India, Egypt, and China. It is burned as fuel in stoves and ovens. The smoke is distasteful and the burning causes the loss of fertilizer nitrogen.

In developed countries, some attempts to use manure from large-scale feedlots have been tried. Wells et al. rolled feedlot manure into dried logs.[6] These did not burn readily, smoldered to an ash, and did not give off much heat. The direct burning of livestock waste for energy has been largely discarded in North America and Europe because of the minimal heating value, the malodors generated during the burning, and the higher value of the waste as a fertilizer for crops.

Livestock wastes often include a large fraction of fiber composed of undigested feed or bedding material used in the housing system.[1] This fibrous matter has absorbed water, and the moisture content of the waste is high. The common function of bedding is to provide cushioning for the animal and to absorb moisture. Some operators have been successful in utilizing livestock wastes for bedding.[7] The fibrous material is separated from the fines and free water by running the livestock wastes over screens, which catch the long fibers and allow the fines and free water to pass through.[8] Then the fibrous fraction is dried by spreading in the sun and wind or by forced aeration. The moisture content of the fibers is often greater than 20%.[7] Considerable energy is required to remove the moisture and produce a satisfactory bedding that will absorb moisture and provide a comfortable environment for the animal. Air drying is suitable for dry climates and for dry periods of the year in some temperate climates. The cost of heated-air drying exceeds the cost of new bedding in almost all climates.

Composting of livestock waste has been used to reduce the moisture content and provide a suitable bedding material. The composting process uses the aerobic microorganisms that heat and stabilize the waste.[1] During this heating process, moisture is released from the waste, and the resultant compost is drier and more stable. Hence it is a much better bedding material than the original livestock waste.

Livestock waste removed from some specialized housing units that use liberal amounts of bedding may be recycled as bedding directly by an operator who has less demanding requirements for bedding. For example, an artifical insemination bull barn may produce livestock waste that can be reused as bedding by a beef feeder using loose housing.

Single-cell protein can be grown on livestock waste.[9-11] The protein is reclaimed from the processed livestock waste and utilized as a supplement in animal feed. Both aerobic and anaerobic treatment processes can be used to harvest single-cell protein. A thermophilic anaerobic fermenter was used to process beef feedlot waste at The Roman L. Hruska Meat Animal Research Center.[9] The overall process was reported to be technically and economically feasible for large-scale operations. More technical development is needed to increase the protein recovery above the 20% that can be recovered by centrifugation.

The fermenter effluent can be mixed directly with grain and supplements. This is an effective method of incorporating the protein into a ration. However, the resultant feed has very high moisture content and may freeze in feedbunks. Bunk life of the feed is reduced during summer temperatures. There is a need for further separation of the high-quality protein from the water and inorganic salts.[9]

The use of oxidation ditches to produce single-cell protein is reported by Day and Harmon.[12] The increased cost of electrical energy has reduced the economic viability of these systems because they consume large amounts of electricity in the motors that operate the aeration wheels. The oxidation ditch also drives large amounts of nitrogen into the atmosphere, and then this nitrogen is not available for land application to crops.

Thermochemical processes can be used to convert livestock wastes to useful products.[4,13] The processes are complex and involve high temperatures, high pressures, and rigid control of operating parameters. The conversion processes are not well developed, and projections show that they may be economically feasible only when implemented on a very large scale. Scale up from research tests to production plants has not occurred. The literature provides detailed reports of the bench-scale experiments that have been conducted on thermochemical processes.[4,13-16] These reports provide some insight into how, someday, synthetic natural gas, ethylene, ammonia synthesis gas, char, and other byproducts may be obtained from livestock manure.[3]

There is mention in the literature of some other byproducts from livestock wastes. A hardboard has been made of the fibrous portion of the livestock waste.[17] The board was not as strong as other commercial hardboards and was not water resistant.

The simple process of harvesting water from a waste management system may be understood as a byproduct recovery system. Some waste management systems use large amounts of water to move the waste from the housing unit to storage.[1] Some systems collect large volumes of rainfall runoff from open feedlots.[1] The water may be cleaned up and then reused. A settling pond can remove solids from flush water and allow the water to be pumped back to the flush system.[7] Here the water is a byproduct because it replaces fresh water. Water pumped onto cropland from treatment systems can replace expensive irrigation water. This is perhaps the most frequently overlooked byproduct of livestock wastes.

In summary, there are many byproducts that can be extracted from livestock wastes. At the current level of technological development, most are not being commercially obtained from livestock wastes. Single-cell protein growing and harvesting is being implemented by a few innovators. Water is being reused in many waste management systems to replace purchased fresh water or fresh water pumped from wells. Other processes are undergoing further development to try and make them feasible. Some processes may never become feasible.

REFERENCES

1. **Miner, R. J. and Smith, R. J., Eds.**, Livestock Waste Management with Pollution Control, North Central Regional Res. Publ. 222, Midwest Plan Service No. 19, Midwest Plan Service, Ames, Iowa, 1970, 7.
2. **Mackenzie, J. D.**, Method of Making Foamed Glass Product with Excretia and Glass Batch, U.S. Patent 3,811,851, 1974 (Assignee: The Regents of the University of California).
3. **Kreis, R. D.**, Recovery of By-Products from Animal Wastes — A Literature Review, EPA-600/2-79-142, U.S. Environmental Protection Agency, Ada, Okla., 1971, 1.
4. **White, R. K. and Taiganides, E. P.**, Pyrolysis of livestock wastes, *Proc. Int. Symp. Livestock Wastes*, Ohio State University, Columbus, 1971, 190.
5. **Davis, E. G., Feld, I. L., and Brown, J. H.**, Combustion Disposal of Manure Wastes and Utilization of the Residue, Bureau of Mines Solid Waste Res. Prog. Tech. Prog. Rep. 46, U.S. Department of the Interior, Washington, D.C., 1972.
6. **Wells, D. M., Whetstone, G. A., and Sweazy, R. M.**, Manure, How It Works, presented at the ANCA-EPA Action Conf., Denver, 1973, 1.
7. **Anon.** Livestock Waste Facilities Handbook, Midwest Plan Service Publ. No. 18, Midwest Plan Service, Ames, Iowa, 1975.
8. **Fairbank, W. C., Bishop, S. E., and Chang, A. C.**, Dairy waste fiber — a byproduct with a future?, in Managing Livestock Wastes, Proc. 3rd Int. Symp. Livestock Wastes, American Society of Agricultural Engineers, St. Joseph, Mich., 1975, 135.
9. **Boersma, L. L., Gasper, E., Miner, J. R., Oldfield, J. E., Phinney, H. K., and Cheeke, P. R.**, Management of Swine Manure for Recovery of Protein and Biogas, Agric. Exp. Stn. Rep. No. 507, Water Resources Research Institute, Oregon State University, Corvallis, 1978.
10. **Hashimoto, A. G., Chen, Y. R., and Prior, R. L.**, Methane and Protein Production from Animal Feedlot Wastes, presented at 33rd Annu. Meet. SCSA, Denver, July 30 to August 2, 1978, 1.
11. **Wallick, J., Harper, J. M., Tenerdy, R. P., and Murphy, V. G.**, Anaerobic Fermentation of Manure into Single Cell Protein, presented at Summer Meeting of American Society of Agricultural Engineers, Logan, Utah, June 27 to 30, 1978, 1.
12. **Day, D. L. and Harmon, B. G.**, A recycled feed source from aerobically processed swine wastes, *Trans. ASAE*, 17, 82, 1974.
13. **Walawender, W. P., Fan, L. T., Engler, C. R., and Erickson, L. E.**, Feedlot Manure and Other Agricultural Wastes as Future Material and Energy Sources. II. Process Description, Rep. No. 45, Institute for Systems Design and Optimization, Kansas State University, Manhattan, 1973, 1.
14. **Walawender, W. P., Fan, L. T., Engler, C. R., and Erickson, L. E.**, Feedlot Manure and Other Agricultural Wastes as Future Material and Energy Sources. III. Economic Evaluations, Contribution No. 33, Department of Chemical Engineering, Kan. Agric. Exp. Stn. Manhattan, 1973, 1.
15. **Huffman, W. J., Halligan, J. E., and Peterson, R. L.**, Conversion of Cattle Feedlot Manure to Ethylene and Ammonia Synthesis Gas, EPA-600/2-78-026, U.S. Environmental Protection Agency, Ada, Okla., 1978, 1.
16. **Halligan, J. E., Herzog, K. L., Parker, W. H., and Sweazy, R. M.**, Conversion of Cattle Feedlot Wastes to Ammonia Synthesis Gas, EPA-660/2-74-090, U.S. Environmental Protection Agency, Ada, Okla., 1974, 1.
17. **Sloneker, J. H., Jones, R. W., Griffin, H. L., Eskins, K., Bucher, B. L., and Inglett, G. E.**, Processing Animal Wastes for Feed and Industrial Products, Symp. Processing Agricultural and Municipal Wastes, New York, 1972, 13.

Appendix

PREFERRED UNITS FOR EXPRESSING PHYSICAL QUANTITIES (AND THE CONVERSION FACTORS)

ASAE Engineering Practice EP285.6

USE OF SI (METRIC) UNITS*

SECTION 1 — PURPOSE AND SCOPE

1.1 This Engineering Practice is intended as a guide for uniformity incorporating the international System of Units (SI). It is intended for use in implementing ASAE policy, "Use of SI Units in ASAE Standards, Engineering Practices, and Data." This Engineering Practice includes a list of preferred units and conversion factors.

SECTION 2 — SI UNITS OF MEASURE

2.1 SI consists of seven base units, two supplementary units, a series of derived units consistent with the base and supplementary units. There is also a series of approved prefixes for the formation of multiples and submultiples of the various units. A number of derived units are listed in paragraph 2.1.3 including those with special names. Additional derived units without special names are formed as needed from base units or other derived units, or both.

2.1.1 Base and supplementary units. For definitions refer to International Organization for Standardization ISO 1000, SI Units and Recommendations for the Use of Their Multiples and of Certain Other Units.

Base Units
  Meter (m) — unit of length
  Second (s) — unit of time
  Kilogram (kg) — unit of mass
  Kelvin (K) — unit of thermodynamic temperature
  Ampere (A) — unit of electric current
  Candela (cd) — luminous intensity
  Mole (mol) — the amount of a substance

Supplementary units
  Radian (rad) — plane angle
  Steradian (sr) — solid angle

2.1.2 SI unit prefixes

Multiples and submultiples	Prefix	SI Symbol
10^{18}	exa	E
10^{15}	peta	P

*   Prepared under the general direction of ASAE Committee on Standards (T-1); reviewed and approved by ASAE division standardizing committees: Power and Machinery Division Technical Committee (PM-03), Structures and Environment Division Technical Committee (SE-03), Electric Power and Processing Division Technical Committee (EPP-03), Soil and Water Division Standards Committee (SW-03), and Education and Research Division Steering Committee; adopted by ASAE December 1964; reconfirmed for one year, December 1969, December 1970, December 1971, December 1972; revised by the Metric Policy Subcommittee December 1973; revised March 1976; revised and reclassified as an Engineering Practice, April 1977; revised April 1979, revised editorially December 1979; revised September 1980, February 1982; revised editorially January 1985.

10^{12}	tera	T
10^9	giga	G
10^6	mega	M
10^3	kilo	k
10^2	hecto	h
10^1	deka	da
10^{-1}	deci	d
10^{-2}	centi	c
10^{-3}	milli	m
10^{-6}	micro	μ
10^{-9}	nano	n
10^{-12}	pico	p
10^{-15}	femto	f
10^{-18}	atto	a

2.1.3 Derived units are combinations of based units or other derived units as needed to describe physical properties, for example, acceleration. Some derived units are given special names; others are expressed in the appropriate combination of SI units. Some currently defined derived units are tabulated in Table 1.

SECTION 3 — RULES FOR SI USAGE

3.1 General. The established SI units (base, supplementary, derived, and combinations thereof with appropriate multiple or submultiple prefixes) should be used as indicated in this section.

3.2 Application of prefixes. The prefixes given in paragraph 2.1.2 should be used to indicate orders of magnitude, thus eliminating insignificant digits and decimals, and providing a convenient substitute for writing powers of 10 as generally preferred in computation. For example:

12 300 m or 12.3×10^3 m becomes 12.3 km, and

0.0123 mA or 12.3×10^{16} A becomes 12.3 μA

It is preferable to apply prefixes to the numerator of compound units, except when using kilogram (kg) in the denominator, since it is a base unit of SI and should be used in preference to the gram. For example:

Use 200 J/kg, not 2 dJ/g

With SI units higher order such as m^2 or m^3, the prefix is also raised to the same order. For example:

mm^2 is $(10^{-3}$ m$)^2$ or 10^{-6} m^2

3.3 Selection of prefix. When expressing a quantity by a numerical value and a unit, a prefix should be chosen so that the numerical value preferably lies between 0.1 and 1000, except where certain multiples and submultiples have been agreed for particular use. The same unit, multiple, or submultiple should be used in tables even though the series may exceed the preferred range of 0.1 to 1000. Double prefixes and hyphenated prefixes should not be used. For example:

use GW (gigawatt) and kMW

3.4 Capitalization. Symbols for SI units are only capitalized when the unit is derived from a proper name; for example, N for Isaac Newton (except liter, L). Unabbreviated units are not capitalized; for example kelvin and newton. Numerical prefixes given in paragraph 2.1.2 and their symbols are not capitalized; except for the symbols M (mega), G (giga), T (tera), P (peta), and E (exa).

3.5 Plurals. Unabbreviated SI units form their plurals in the usual manner. SI symbols are

Table 1
DERIVED UNITS

Quantity	Unit	SI Symbol	Formula
Acceleration	Meter per second squared	—	m/s^2
Activity (of a radioactive source)	Disintegration per second	—	(disintegration)/s
Angular acceleration	Radian per second squared	—	rad/s^2
Angular velocity	Radian per second	—	rad/s
Area	Square meter	—	m^2
Density	Kilogram per cubic meter	—	kg/m^3
Electrical capacitance	Farad	F	A·s/V
Electrical conductance	Siemens	S	A/V
Electrical field strength	Volt per meter	—	V/m
Electrical inductance	Henry	H	V·s/A
Electrical potential difference	Volt	V	W/A
Electrical resistance	Ohm	Ω	V/A
Electromotive force	Volt	V	W/A
Energy	Joule	J	N·m
Entropy	Joule per kelvin	—	J/K
Force	Newton	N	$kg·m/s^2$
Frequency	Hertz	Hz	(cycle)/s
Illuminance	Lux	lx	lm/m^2
Luminance	Candela per square meter	—	cd/m^2
Luminous flux	Lumen	lm	cd·sr
Magnetic field strength	Ampere per meter	—	A/m
Magnetic flux	Weber	Wb	V·s
Magnetic flux density	Tesla	T	Wb/m^2
Magnetomotive force	Ampere	A	—
Power	Watt	W	J/s
Pressure	Pascal	Pa	N/m^2
Quantity of electricity	Coulomb	C	A·s
Quantity of heat	Joule	J	N·m
Radiant intensity	Watt per steradian	—	W/sr
Specific heat	Joule per kilogram-kelvin	—	J/kg·K
Stress	Pascal	Pa	N/m^2
Thermal conductivity	Watt per meter-kelvin	—	W/m·K
Velocity	Meter per second	—	m/s
Viscosity, dynamic	Pascal-second	—	Pa·s
Viscosity, kinematic	Square meter per second	—	m^2/s
Voltage	Volt	V	W/A
Volume	Cubic meter	—	m^3
Wavenumber	Reciprocal meter	—	(wave)/m
Work	Joule	J	N·m

always written in singular form. For example:

50 newtons or 50 N

25 millimeters or 25 mm

3.6   Punctuation. Whenever a numerical value is less than one, a zero should precede the decimal point. Periods are not used after any SI unit symbol, except at the end of a sentence. English speaking countries use a dot for the decimal point, others use a comma. Use spaces instead of commas for grouping numbers into threes (thousands). For example:

6 357 831.376 88

not 6,357,831.367,88

3.7   Derived units. The product of two or more units in symbolic form is preferably indicated by a dot midway in relation to unit symbol height. The dot may be dispensed with when there is no risk of confusion with another unit symbol. For example:

Use N · m or N m, but not mN

A solidus (oblique stroke, /) a horizontal line, or negative powers may be used to express a derived unit formed from two others by division. For example:

m/s, $\dfrac{m}{s}$, or m · s^{-1}

Only one solidus should be used in a combination of units unless parentheses are used to avoid ambiguity.

3.8   Representation of SI units in systems with limited character sets. For computer printers and other systems which do not have the characters available to print SI units correctly, the methods shown in ISO 2955, Information Processing — Representation of SI and Other Units for Use in Symbols with Limited Character Sets, is recommended.

SECTION 4 — NON-SI UNITS

4.1   Certain units outside the SI are recognized by ISO because of their practical importance in specialized fields. These include units for temperature, time, and angle. Also included are names for some multiples of units such as "liter" (L)* for volume, "hectare" (ha) for land measure and "metric tone" (t) for mass.

4.2   Temperature. The SI base unit for thermodynamic temperature is kelvin (K). Because of the wide usage of the degree Celsius, particularly in engineering and nonscientific areas, the Celsius scale (formerly called the centigrade scale) may be used when expressing temperature. The Celsius scale is related directly to the kelvin scale as follows:

one degree Celsius (1°C) equals one kelvin (1 K), exactly

A Celsius temperature (t) is related to a kelvin temperature (T), as follows:

t = T − 273.15

4.3   Time. The SI unit for time is the second. This unit is preferred and should be used when technical calculations are involved. In other cases use of the minute (min), hour (h), day (d), etc. is permissible.

4.4   Angles. The SI unit for plane angle is the radian. The use of arc degrees (°) and its decimal or minute ('), second (") submultiples is permissible when the radian is not a convenient unit. Solid angles should be expressed in steradians.

SECTION 5 — PREFERRED UNITS AND CONVERSION FACTORS

5.1   Preferred units for expressing physical quantities commonly encountered in agricultural engineering work are listed in Table 2. These are presented as an aid to selecting proper units for given applications and to promote consistency when interpretation of the general rules of SI may not produce consistent results. Factors for conversion from old units to SI units are included in Table 2.

SECTION 6 — CONVERSION TECHNIQUES

6.1   Conversion of quantities between systems of units involves careful determination of the number of significant digits to be retained. To convert "1 quart of oil" to "0.946 352 9 liter of oil" is, of course, unrealistic because the intended accuracy of the value does not warrant expressing the conversion in this fashion.

---

*   The International symbol for liter is either the lowercase "l" or the uppercase "L". ASAE recommends the use of uppercase "L" to avoid confusion with the numeral "1".

Table 2
PREFERRED UNITS FOR EXPRESSING PHYSICAL QUANTITIES

Quantity	Application	From: old units	To: SI units	Multiply by:
Acceleration, angular	General	rad/s²	rad/s²	
Acceleration, linear	Vehicle	(mile/h)/s	(km/h)/s	1.609 344ᵃ
	General (includes acceleration of gravity)ᵇ	ft/s²	m/s²	0.304 8ᵃ
Angle, plane	Rotational calculations	r (revolution)	r (revolution)	
		rad	rad	
	Geometric and general	° (deg)	°	
		' (min)	° (decimalized)	1/60ᵃ
		' (min)	'	
		" (sec)	° (decimalized)	1/3600ᵃ
		" (sec)	"	
Angle, solid	Illumination calculations	sr	sr	
Area	Cargo platforms, roof and floor area, frontal areas, fabrics, general	in.²	m²	0.000 645 16ᵃ
		ft²	m²	0.092 903 04ᵃ
	Pipe, conduit	in.²	mm²	645.16ᵃ
		in.²	cm²	6.451 6ᵃ
	Small areas, orifices, cross section area of structural shapes	ft²	m²	0.092 903 04ᵃ
		in.²	mm²	645.16ᵃ
	Brake and clutch contact area, glass, radiators, feed opening	in.²	cm²	6.451 6ᵃ
	Land, pond, lake, reservoir, open water channel (small)	ft²	m²	0.092 903 04ᵃ
	(large)	acre	ha	0.404 687 3(d)
	(very large)	mile²	km²	2.589 998
Area per time	Field operations	acre/h	ha/h	0.404 687 3
	Auger sweeps, silo unloader	ft²/s	m²/s	0.092 903 04ᵃ
Bending moment	(See moment of force)			
Capacitance, electric	Capacitors	μF	μF	
Capacity, electric	Battery rating	A·h	A·h	
Capacity, heat	General	Btu/°f	kJ/Kᶜ	1.899 101
Capacity, heat, specific	General	Btu/(lb·°F)ᶜ	kJ/(kg·K)ᶜ	4.186 8ᵃ

Table 2 (continued)

PREFERRED UNITS FOR EXPRESSING PHYSICAL QUANTITIES

Quantity	Application	From: old units	To: SI units	Multiply by:
Capacity, volume	(See volume)			
Coefficient of heat transfer	General	Btu/(h·ft²·°F)	W/(m²·K)[c]	5.678 263
Coefficient of linear expansion	Shrink fit, general	°F⁻¹, (1/°F)	K⁻¹, (1/K)[c]	1.8[a]
Conductance, electric	General	mho	S	1[a]
Conductance, thermal	(See coefficient of heat transfer)			
Conductivity, electric	Material property	mho/ft	S/m	3.280 840
Conductivity, thermal	General	Btu·ft/(h·ft²·°F)	W/(m·K)[c]	1.730 735
Consumption, fuel	Off highway vehicles (see also efficiency, fuel)	gal/h	L/h	3.785 412
Consumption, oil	Vehicle performance testing	qt/(1000 miles)	L/(1000 km)	0.588 036 4
Consumption, specific, oil	Engine testing	lb/(hp·h)	g/(kW·h)	608.277 4
		lb/(hp·h)	g/MJ	168.965 9
Current, electric	General	A	A	
Density, current	General	A/in.²	kA/m²	1.550 003
		A/ft²	A/m²	10.763 91
Density, magnetic flux	General	Kilogauss	T	0.1[a]
Density, (mass)	Solid, general; agricultural products, soil, building materials	lb/yd³	kg/m³	0.593 276 3
		lb/in.³	kg/m³	27 679.90
		lb/ft³	kg/m³	16.018 46
	Liquid	lb/gal	kg/L	0.119 826 4
	Gas	lb/ft³	kg/m³	16.018 46
	Solution concentration	—	g/m³, mg/L	—
Density of heat flow rate	Irradiance, general	Btu/(h·ft²)	W/m²	3.154 591[d]
Consumption, fuel	(See flow, volume)			
Consumption, specific fuel	(See efficiency, fuel)			
Drag	(See force)			
Economy, fuel	(See efficiency, fuel)			

Quantity	Application	To convert from	to	Multiply by
Efficiency, fuel				
Highway vehicles				
Economy		mile/gal	km/L	0.415 143 7
Consumption		—	L/(100 km)	[c]
Specific fuel consumption		lb/(hp·h)	g/MJ	168.965 9
Off-highway vehicles				
Economy		hp·h/gal	kW·h/L	0.196 993 1
Specific fuel consumption		lb/(hp·h)	g/MJ	168.965 9
Specific fuel consumption		lb/(hp·h)	kg/(kW·h)[f]	0.608 277 4
Energy, work, enthalpy, quantity of heat				
Impact strength		ft·lbf	J	1.355 818
Heat		Btu	kJ	1.055 056
		kcal	kJ	4.186 8[a]
Energy, usage, electrical		kW·h	kW·h	
		kW·h	MJ	3.6
Mechanical, hydraulic, general		ft·lbf	J	1.355 818
		ft·pdl	J	0.042 140 11
		hp·h	MJ	2.684 520
		hp·h	kW·h	0.745 699 9
Energy per area				
Solar radiation		Btu/ft^2	MJ/m^2	0.011 356 528
Energy, specific				
General		cal/g[g]	J/g	4.186 8[a]
		Btu/lb	kJ/kg	2.326[a]
Enthalpy	(See energy)			
Entropy	(See capacity, heat)			
Entropy, specific	(See capacity, heat, specific)			
Floor loading	(See mass per area)			
Flow, heat, rate	(See power)			
Flow, mass, rate	Gas, liquid	lb/min	kg/min	0.453 592 4
		lb/s	kg/s	0.453 592 4
	Dust flow	g/min	g/min	
	Machine work capacity, harvesting, materials handling	ton (short)/h	t/h, Mg/h[a]	0.907 184 7
Flow, volume	Air, gas, general	ft^3/s	m^3/s	0.028 316 85
		ft^3/s	m^3/min	1.699 011
	Liquid flow, general	gal/s (gps)	L/s	3.785 412
		gal/s (gps)	m^3/s	0.003 785 412
		gal/min (gpm)	L/min	3.785 412
		oz/min	mL/s	29.573 53
	Seal and packing leakage, sprayer flow	oz/min	mL/min	29.573 53

Table 2 (continued)
PREFERRED UNITS FOR EXPRESSING PHYSICAL QUANTITIES

Quantity	Application	From: old units	To: SI units	Multiply by:
	Fuel consumption	gal/h	L/h	3.785 412
	Pump capacity, coolant flow, oil flow	gal/min (gpm)	L/min	3.785 412
	Irrigation sprinkler, small pipe flow	gal/min (gpm)	L/s	0.063 090 20
	River and channel flow	ft³/s	m³/s	0.028 316 85
Flux, luminous	Light bulbs	lm	lm	
Flux, magnetic	Coil rating	maxwell	Wb	0.000 000 01ª
Force, thrust, drag	Pedal, spring, belt, hand lever, general	lbf	N	4.448 222
		ozf	N	0.278 013 9
		pdl	N	0.138 255 0
		kgf	N	9.806 650
		dyne	N	0.000 01ª
	Drawbar, breakout, rim pull, winch line pull,ʰ general	lbf	kN	0.004 448 222
Force per length	Beam loading	lbf/ft	N/m	14.593 90
	Spring rate	lbf/in.	N/mm	0.175 126 8
Frequency	System, sound and electrical	Mc/s	MHz	1ª
		kc/s	kHz	1ª
		Hz, c/s	Hz	1ª
	Mechanical events, rotational	r/s (rps)	s⁻¹, r/s	1ª
	Engine, power-take-off shaft, gear speed	r/min (rpm)	min⁻¹, r/min	1ª
	Rotational dynamics	r/min (rpm)	min⁻¹, r/min	1ª
		rad/s	rad/s	
Hardness	Conventional hardness numbers, BHN, R, etc. not affected by change to SI			
Heat	(See energy)			
Heat capacity	(See capacity, heat)			
Heat capacity, specific	(See capacity, heat, specific)			
Heat flow rate	(See power)			
Heat flow — density of	(See density of heat flow)			
Heat, specific	General	cal/g·°C	kJ/kg·K	4.186 8ª
		Btu/lb·°F	kJ/kg·K	4.186 8ª

Quantity	Application	From	To	Multiply by
Heat transfer coefficient	(See coefficient of heat transfer)			
Illuminance, illumination	General	fc	lx	10.763 91
Impact strength	(See strength, impact)			
Impedance, mechanical	Damping coefficient	lbf·s/ft	N·s/m	14.593 90
Inductance, electric	Filters and chokes, permeance	H	H	
Intensity, luminous	Light bulbs	candlepower	cd	1[a]
Intensity, radiant	General	W/sr	W/sr	
Leakage	(See flow, volume)			
Length	Land distances, maps, odometers	mile	km	1.609 344[a,h]
	Field size, turning circle, braking distance, cargo platforms, rolling circumference, water depth, land leveling (cut and fill)	rod	m	5.029 210[h]
		yd	m	0.914 4
		ft	m	0.304 8[a]
	Row spacing	in.	cm	2.54[a]
	Engineering drawings, product specifications, vehicle dimensions, width of cut, shipping dimensions, digging depth, cross section of lumber, radius of gyration, deflection	in.	mm	25.4[a]
	Precipitation, liquid, daily and seasonal, field drainage (runoff), evaporation and irrigation depth	in.	mm	25.4[a]
	Precipitation, snow depth	in.	cm	2.54[a]
	Coating thickness, filter particle size	mil	μm	25.4[a]
		μin.	μm	0.025 4[a]
		micron	μm	1[a]
	Surface texture			
	Roughness, average	μin.	μm	0.025 4[a]
	Roughness sampling length, waviness height and spacing	in.	mm	25.4[a]
	Radiation wavelengths, optical measurements (interference)	μin.	nm	25.4[a]
Length per time	Precipitation, liquid per hour	in./h	mm/h	25.4[a]
	Precipitation, snow depth per hour	in.h	cm/h	2.54[a]
Load	(See mass)			
Luminance	Brightness	footlambert	cd/m²	3.426 259
Magnetization	Coil field strength	A/in.	A/m	39.370 08

Table 2 (continued)

PREFERRED UNITS FOR EXPRESSING PHYSICAL QUANTITIES

Quantity	Application	From: old units	To: SI units	Multiply by:
Mass	Vehicle mass, axle rating, rated load, tire load, lifting capacity,¹ tipping load, load, quantity of crop, counter mass, body mass	ton (long)	t, Mg¹	1.016 047
		ton (short)	t, Mg¹	0.907 184 7
	general	lb	kg	0.453 592 4
		slug	kg	14.593 90
Small mass		oz	g	28.349 52
Mass per area	Fabric, surface coatings	oz/yd²	g/m²	33.905 75
		lb/ft²	kg/m²	4.882 428
		oz/ft²	g/m²	305.151 7
	Floor loading	lb/ft²	kg/m²	4.882 428
	Application rate, fertilizer, pesticide	lb/acre	kg/ha	1.120 851
	Crop yield, soil erosion	ton (short)/acre	t/ha¹	2.241 702
Mass per length	General, structural members	lb/ft	kg/m	1.488 164
		lb/yd	kg/m	0.496 054 7
Mass per time	Machine work capacity, harvesting, materials handling	ton (short)/h	t/h, Mg/h¹	0.907 184 7
Modulus of elasticity Modulus of rigidity	General	lbf/in.²	MPa	0.006 894 757
Modulus, section	(See modulus of elasticity) General	in.³	mm³	16.387.06
		in.³	cm³	16.387 06
Modulus, bulk	System fluid compression	psi	kPa	6.894 757
Moment, bending	(See moment of force)			
Moment of area, second	General	in.⁴	mm⁴	416 231.4
		in.⁴	cm⁴	41.623 14
Moment of force, torque, bending moment	General, engine torque, fasteners, steering torque, gear torque, shaft torque	lbf·in.	N·m	0.112 984 8
		lbf·ft	N·m	1.355 818
		kgf·cm	N·m	0.098 066 5ᵃ
Moment of inertia, mass	Locks, light torque	ozf·in.	mN·m	7.061 552
Moment of mass	Flywheel, general	lb·ft²	kg·m²	0.042 140 11
Moment of momentum	Unbalance	oz·in.	g·m	0.720 077 8
	(See momentum, angular)			

Quantity	Description	From	To	Factor
Moment of section	(See moment of area, second)			
Momentum, linear	General	lbf·ft/s	kg·m/s	0.138 255 0
Momentum, angular	Orsional vibration	lbf·ft²/s	kg·m²/s	0.042 140 11
Permeability	Magnetic core properties	H/ft	H/m	3.280 840
Permeance	(See inductance)			
Potential, electric	General	V	V	
Power	General, light bulbs	W	W	
Power	Air conditioning, heating	Btu/min	W	17.584 17
		Btu/h	W	0.293 071 1
	Engine, alternator, drawbar, power take-off, hydraulic and pneumatic systems, heat rejection, heat exchanger capacity, water power, electrical power, body heat loss	hp (550 ft·lbf/s)	kW	0.745 699 9
Power per area	Solar radiation	Btu/ft²·h	W/m²	3.154 591
Pressure	All pressures except very small	lbf/in.² (psi)	kPa	6.894 757
		in.Hg (60°F)	kPa	3.376 85
		in.H₂O (60°F)	kPa	0.248 84
		mmHg (0°C)	kPa	0.133 322
		kgf/cm²	kPa	98.066 5
		bar	kPa	100.0ª
		lbf/ft²	kPa	0.047 880 26
		atm (normal = 760 torr)	kPa	101.325ª
Pressure, sound level	Very small pressures (high vacuum)	lbf/in.² (psi)	Pa	6 894.757
	Acoustical measurement — when weighting is specified show weighting level in parenthesis following the symbol, for example dB(A)	dB	dB	
Quantity of electricity	General	C	C	
Radiant intensity	(See intensity, radiant)			
Resistance, electric	General	Ω	Ω	
Resistivity, electric	General	Ω·ft	Ω·m	0.304 8ª
		Ω·ft	Ω·cm	30.48ª
Sound pressure level	(See pressure, sound, level)			
Speed	(See velocity)			
Spring rate, linear	(See force per length)			
Spring rate, torsional	General	lbf·ft/deg	N·m/deg	1.355 818

Table 2 (continued)
PREFERRED UNITS FOR EXPRESSING PHYSICAL QUANTITIES

Quantity	Application	From: old units	To: SI units	Multiply by:
Strength, field, electric	General	V/ft	V/m	3.280 840
Strength, field, magnetic	General	Oersted	A/m	79,577 47
Strength, impact	Materials testing	ft·lbf	J	1.355 818
Stress	General	lbf/in.^{2}	MPa	0.006 894 757
Surface tension	(See tension, surface)			
Temperature	General use	°F	°C	$t_C = (t_F - 32)/1.8^a$
	Absolute temperature, thermodynamics, gas cycles	°R	K	$T_K = T_R/1.8^a$
Temperature interval	General use	°F	K^c	1 K = 1°C = 1.8°F^a
Tension, surface	General	lbf/in.	mN/m	175 126.8
		dyne/cm	mN/m	1^a
Thermal diffusivity	Heat transfer	ft^2/h	m^2/h	0.092 903 04
Thrust	(See force)			
Time	General	s	s	
		h	h	
		min	min	
	Hydraulic cycle time	s	s	
	Hauling cycle time	min	min	
Torque	(See moment of force)			
Toughness, fracture	Metal properties	ksi·in.$^{0.5}$	MPa·m$^{0.5}$	1.098 843
Vacuum	(See pressure)			
Velocity, angular	(See velocity, rotational)			
Velocity, linear	Vehicle	mile/h	km/h	1.609 344^a
	Fluid flow, conveyor speed, lift speed, air speed	ft/s	m/s	0.304 8^a
	Cylinder actuator speed	in./s	mm/s	25.4^a
	General	ft/s	m/s	0.304 8^a
		ft/min	m/min	0.304 8^a
		in./s	mm/s	25.4^a

Quantity	Application/Description	From	To	Factor
Velocity, rotational	(See frequency)			
Viscosity, dynamic	General liquids	Centipoise	mPa·s	1[a]
Viscosity, kinematic	General liquids	Centistokes	mm²/s	1[a]
Volume	Truck body, shipping or freight, bucket capacity, earth, gas, lumber, building, general	yd³	m³	0.764 554 9
		ft³	m³	0.028 316 85
	Combine harvester grain tank capacity	Bushel	L	35.239 07
	Automobile luggage capacity	ft³	L	28.316 85
	Gas pump displacement, air compressor, air reservoir, engine displacement			
	Large	in.³	L	0.016 387 06
	Small	in.³	cm³	16.387 06
	Liquid — fuel, lubricant, coolant, liquid	gal	L	3.785 412
		qt	L	0.946 352 9
		pt	L	0.473 176 5
	wheel ballast	pt	L	0.473 176 5
	Small quantity liquid	oz	mL	29.573 53
	Irrigation, reservoir	acre·ft	m³	1 233.489[h]
			dam³	1.233 489[h]
	Grain bins	bushel (U.S.)	m³	0.035 239 07
Volume per area	Application rate, pesticide	gal/acre	L/ha	9.353 958
Volume per time	Fuel consumption (also see flow)	gal/h	L/h	3.785 412
Weight	May mean either mass or force — avoid use of weight			
Work	(See energy)			
Young's modulus	(See modulus of elasticity)			

Notes: 1. Quantities are arranged in alphabetical order by principal nouns. For example, surface tension is listed as tension, surface.

2. All possible applications are not listed, but others such as rates can be readily derived. For example, from the preferred units for energy and volume the units for heat energy per unit volume, kJ/m³, may not be derived.

3. Conversion factors are shown to seven significant digits, unless the precision with which the factor is known does not warrant seven digits.

a   Indicates exact conversion factor.

b   Standard acceleration of gravity is 9.806 650 m/s² exactly (Adopted by the General Conference on Weights and Measures).

c   In these expressions, K indicates temperature intervals. Therefore K may be replaced with °C if desired without changing the value or affecting the conversion factor. kJ/(kg·K) = kJ/(kg·°C).

d   Conversions of Btu are based on the International Table Btu.

e   Convenient conversion: 235.215 ÷ (mile per gal) = L/(100 km).

Table 2 (continued)
PREFERRED UNITS FOR EXPRESSING PHYSICAL QUANTITIES

f ASAE S209 and SAE J708, Agricultural Tractor Test Code, specify kg/(kW·h). It should be noted that there is a trend toward use of g/MJ as specified for highway vehicles.

g Not to be confused with kcal/g. kcal often called calorie.

h Official use in surveys and cartography involves the U.S. survey foot, which is longer than the international foot by two parts per million. The factors used in this standard for acre, acre foot, rod are based on the U.S. survey foot. Factors for all other old length units are based on the international foot. (See ANSI/ASTM Standard E380-76, Metric Practice).

i Lift capacity ratings for cranes, hoists, and related components such as ropes, cable chains, etc. should be rated in mass units. Those items such as winches, which can be used for pulling as well as lifting, shall be rated in both force and mass units for safety reasons.

j The symbol t is used to designate metric ton. The unit metric ton (exactly 1 MG) is in wide use but should be limited to commercial description of vehicle mass, freight mass, and agricultural commodities. No prefix is permitted.

6.2 All conversions, to be logically established, must depend upon an intended precision of the original quantity — either implied by a specific tolerance, or by the nature of the quantity. The first step in conversion is to establish this precision.

6.3 The implied precision of a value should relate to the number of significant digits shown. The implied precision is plus or minus one half unit of the last significant digit in which the value is stated. This is true because it may be assumed to have been rounded from a greater number of digits, and one half of the last significant digit retained is the limit of error resulting from rounding. For example, the number 2.14 may have been rounded from any number between 2.135 and 2.145. Whether rounded or not, a quantity should always be expressed with this implication of precision in mind. For instance, 2.14 in. implies a precision of ±0.005 in., since the last significant digit is in units of 0.01 in.

6.4 Quantities should be expressed in digits which are intended to be significant. The dimension 1.1875 in. may be a very accurate one in which the digit in the fourth place is significant, or it may in some cases be an exact decimalization of a fractional dimension, 1 3/16 in., in which case the dimension is given with too many decimal places relative to its intended precision.

6.5 Quantities should not be expressed with significant zeros omitted. The dimension 2 in. may mean "about 2 in.," or it may, in fact, mean a very accurate expression which should be written 2.0000 in. In the latter case, while the added zeros are not significant in establishing the value, they are very significant in expressing the proper intended precision.

SECTION 7 — RULES FOR ROUNDING

7.1 Where feasible, the rounding of SI equivalents should be in reasonable, convenient, whole units.

7.2 Interchangeability of parts, functionally, physically, or both, is dependent upon the degree of round-off accuracy used in the conversion of the U.S. customary to SI value. American National Standards Institute ANSI/ASTM E380-76, Metric Practice, outlines methods to assure interchangeability.

7.3 Rounding numbers. When a number is to be rounded to fewer decimal places the procedure shall be as follows:

7.3.1 When the first digit discarded is less than 5, the last digit retained shall not be changed. For example, 3.463 25, if rounded to three decimal places, would be 3.463; if rounded to two decimal places, would be 3.46.

7.3.2 When the first digit discarded is greater than 5, or it is a 5 followed by at least one digit other than 0, the last figure retained shall be increased by one unit. For example, 8.376 52, if rounded to three decimal places, would be 8.377; if rounded to two decimal places, would be 8.38.

7.3.3 Round to closest even number when first digit discarded is 5, followed only by zeros.

7.3.4 Numbers are rounded directly to the nearest value having the desired number of decimal places. Rounding must not be done in successive steps to less places. For example:

27.46 rounded to a whole number = 27. This is correct because the "0.46" is less than one half. 27.46 rounded to one decimal place is 27.5. This is correct value. But, if the 27.5 is in turn rounded to a whole number, this is successive rounding and the result, 28, is incorrect.

7.4 Inch-millimeter linear dimensioning conversion. 1 inch (in.) = 25.4 millimeters (mm) exactly. The term "exactly" has been used with all exact conversion factors. Conversion factors not so labeled have been rounded in accordance with these rounding procedures. To maintain intended precision during conversion without retaining an unnecessary number of digits, the millimeter equivalent shall be carried to one decimal place more than the inch value being converted and then rounded to the appropriate significant figure in the last decimal place.

CITED STANDARDS

ASAE S209, Agricultural Tractor Test Code

ANSI/ASTM E380-76, Metric Practice

ISO 1000, SI Units and Recommendations for the Use of Their Multiples and of Certain Other Units

ISO 2955, Information Processing — Representation of SI and Other Units for Use in Systems with Limited Character Sets

Index

INDEX

A